D1304562

A VALENCY PRIMER

A VALENCY PRIMER

J. C. SPEAKMAN, D.Sc.
Reader in Physical Chemistry
University of Glasgow

London
Edward Arnold (Publishers) Ltd.

Boards edition SBN: 7131 2183 1
Paper edition   SBN: 7131 2184 X

Set on Monophoto Filmsetter and printed by
J. W. Arrowsmith Ltd., Bristol, England

Preface

The first version of my small book, *An Introduction to the Modern Theory of Valency*, was written in 1933–4. At that time a first, qualitative electronic interpretation of valency—though hardly new (Sidgwick's great monograph had been published in 1927)—was just beginning to make itself effective in the teaching of chemistry. Since then great developments have occurred. They arise notably from the application of wave mechanics to molecules, and from the gradually accumulating (and now vast) body of information on the sizes, shapes and other intimate properties of molecules. I attempted to keep not too far behind these advances by revising the book in 1943, and again more extensively in 1955, when the indiscreet word 'modern' was deleted from the title. These revisions resulted in a 50 per cent increase in the length of the book.

But, as Sir Cyril Hinshelwood reflected, one may 'doubt whether progress in a subject is reflected only in the increasing size of the books written about it. . . . In some ways the ideal would be that successive editions of a book should get smaller and smaller'. Venturing to follow this counsel, I have now written a much shorter book, only about three-quarters the length of the original. Such curtailment seems justifiable for other reasons besides the fuller understanding we now have of molecules. One reason is that electronic ideas are now much more familiar than they were in the early 1930's. School pupils take them in their stride. No doubt, the same will soon be true—if it be not already—of the wave theory of matter.

Apart from its last chapter, which is taken almost unchanged from the 1955 edition, the text is wholly new. The treatment of elementary electronic ideas is much more concise. More emphasis is laid on the molecular concept, and on the measurable properties of molecules. A primary objective is to develop in the student 'a feeling' for molecules. As George Wald has written, 'if (students) can reach the point of saying to themselves, when up against some problem of molecular behaviour, "What would I do if I were that molecule?", then things are going well'.

Primarily the new book is intended for use in advanced classes in schools and in first-year classes at colleges and universities. Mathematics has been kept at a respectful distance, so that the treatment of wave mechanics has had to be descriptive and naive. Those who pursue the study of chemistry beyond first-year level will need to tackle more rigorous treatments in any case. My aim has been to give the reader a sound, if elementary, basis upon which he may develop a concept of the molecule in chemistry.

The production of the book, in its original edition, owed almost everything to Dr. G. M. Bennett, whose teaching and enthusiasm, at the University of Sheffield in the days when the electronic theory was young, were a source of great inspiration. I should like to think that there are still discernible in this new book some effects of that teaching and that enthusiasm.

April, 1967   J.C.S.

Contents

Chapter 1

The Molecule and Valency

1.1 The atom and the molecule

Matter—for the chemist—consists of atoms. To be sure, the atoms themselves are built up from various fundamental particles, such as electrons, protons and neutrons; but that is rather the concern of the physicist. The chemist usually deals with matter in the form of aggregations of atoms. A number of atoms held together by relatively strong forces, and interacting with other atoms only *via* weak forces (or not at all), is known as a *molecule*. In this book we are mainly concerned with molecules, with their properties—notably their sizes and shapes—and with the forces that hold them together and enable them to exist as recognizable and discrete entities.

As a philosophical concept, the notion of atoms is an ancient one. It began to play an effective part in chemistry only when it could be related to measurable quantities. The atomic theory was applied quantitatively by Dalton (1766–1844) who used it to rationalize his laws of chemical composition. Water, for example, always consists of hydrogen and oxygen combined in definite proportions, which are approximately 1:8 parts by weight. The simplest way of explaining this is to suppose that water is formed by the union of an atom of hydrogen and an atom of oxygen, and that the atoms of hydrogen always have the same relative weight of (say) 1, whilst those of oxygen always have a relative weight of 8. So Dalton represented the composition of water by symbols equivalent to our modern HO.

Later, various chemical arguments led to the conclusion that the relative weights of the atoms were H:O = 1:16, but that two atoms of hydrogen were involved. Water came to be represented by the formula H_2O. There was indeed no reason in strict logic why H_2O should be preferred to (say) H_4O_2, but the former was accepted because it was simpler and led to no chemical inconsistencies. We now have access to more direct evidence which makes us feel certain that H_2O is correct.

For a time there was much confusion between the atom and what we should now call a molecule. The distinction gradually came to be recognized. Atoms are the ultimate chemical units. The working unit of the chemist is usually a collection of atoms—the molecule: H_2 for hydrogen, O_2 for oxygen, H_2O for water, $C_2H_4O_2$ for acetic acid, or $C_6H_{12}O_6$ for glucose. In special cases a molecule may consist of only a single atom. The inert gases have monatomic molecules such as He and Ar; and in the atmosphere of the sun hydrogen is dissociated into single atoms, so that in that environment H may properly be regarded as the molecule.

1.2   Development of the molecular concept

The development of the concept of a molecule since 1811 was much more complicated than can be suggested in a short summary.* A constantly recurring pattern of ideas runs as follows: to explain the observed behaviour of materials—their chemical properties in particular—molecules are postulated with certain properties; these postulates are put forward very tentatively by their originators, whilst they meet objection from those who regard them as too speculative; nevertheless they prove useful in leading to predictions which can be verified; and, if they survive this necessary test, they gradually come to be generally accepted.

The scepticism of the objectors is understandable. All the evidence for molecules was originally indirect; it was mostly based on the chemical compositions of quantities of material containing an enormous—and unknown—number of actual molecules. The constant composition of water was consistent with its consisting of H_2O molecules. But certainly the chemical evidence does not *prove* this. Frankland, for instance, considered it necessary, in 1866, to defend his chemical formulae in these words: 'I am aware that graphic . . . formulae may be objected to, on the ground that students, even when warned against any such interpretation, will be liable to regard them as representations of the actual physical position of the atoms of the compound. In practice I have never found this evil to exist.' Even as late as 1907 the great physical chemist, Ostwald, wrote a textbook on general chemistry without mentioning atoms, or molecules, or structure except in a derisory way (he moderated these extreme views before his death in 1932).

In recent years more direct, physical methods for studying molecules have become available. The striking thing is that they have, in general, fully confirmed the earlier chemical ideas about molecules, and added to them a wealth of circumstantial detail. Water was

* These developments are excellently presented in W. G. Palmer's *Valency, Classical and Modern* (Cambridge University Press, 1959).

supposed to consist of hypothetical H_2O molecules; and Frankland—after cautioning his readers, as we have seen—represented them by the formula $\oplus\!-\!\circledcirc\!-\!\oplus$, which we should now write as (1). We have plenty of physical evidence that this molecule really is triatomic, with the atoms linked in the order shown; that it is bent, rather than straight, with an angle of about 105° at the oxygen atom; and that the distances between this atom and each of the hydrogens are equal and about $1\cdot0 \times 10^{-8}$ cm.

$$H\!-\!O\!-\!H \qquad (1)$$

$$
\begin{array}{c}
H \qquad\quad O \\
| \qquad\ \ \ \diagup\diagup \\
H\!-\!C\!-\!C \qquad\qquad (2)\\
| \qquad\ \ \diagdown \\
H \qquad\quad O\!-\!H
\end{array}
$$

1.3   Molecular and structural formulae

The first stage of the chemical approach was the allocation to an element or compound of its correct *molecular formula*: H_2O, or $C_2H_4O_2$, or $C_6H_{12}O_6$. Nothing was known about the arrangement of these atoms within the supposed molecule, or indeed whether there was any particular arrangement. Then the discovery of iso-merism suggested that the arrangement was specific. Cases were established where two or more different compounds had the same molecular formula: $C_6H_{12}O_6$ represented several sugars besides glucose; $C_2H_4O_2$ represents methyl formate as well as acetic acid. There must therefore be at least two distinct ways of arranging the set of atoms $C_2H_4O_2$.

The next stage which lasted for many decades—and which is still with us so far as work on complex materials or large molecules is concerned—was the deduction of the *structural formula*. This was achieved by a detailed study of the chemical properties of the com-pound (and related compounds), and by application to the results of a set of rules based on experience and governed by strict logic. For example, one of the hydrogen atoms of the acetic acid molecule differs from the other three in that it alone can be replaced easily by sodium to yield sodium acetate, $C_2H_3O_2Na$. It follows that the formula $C_2H_3O_2\!-\!H$ is more informative than $C_2H_4O_2$. From a detailed study of acetic acid in its total chemical context it became possible to represent its reactions by formula (2). The lines represent lines of union, or *bonds*, between the atoms so linked.

This chemical approach proved particularly successful in organic chemistry. By logic based on detailed and often subtle chemical observations, structural formulae were gradually assigned to larger and more complex molecules. The structural theory was one of the most powerful tools ever to be applied in exact science.

1.4   Valency

The word *valency* ('valence' in the United States) became associated with the lines in these structural formulae. It is now used in two rather different senses: first, to describe the type, or types, of *force* existing between atoms in a molecule; secondly, to denote the *number* of bonds, or lines, radiating from a particular atom in a structural formula. This latter was indeed the original meaning, since nothing was then known about the former aspect. Nowadays we know a little about interatomic forces, and they are a principal concern of this book. Valency-as-a-number was sometimes known as 'quanti-valence'. This aspect, as it has now developed, is discussed in more detail in the final chapter.

As information bearing on the structures of molecules gradually accumulated, the importance of valency as a number became evident. A carbon atom always had four bonds, nitrogen usually three, oxygen almost always two, and hydrogen one. Carbon was said to be quadrivalent, nitrogen ter- (or tri-) valent, oxygen bivalent, and hydrogen univalent.* These principles were then of the greatest value in constructing formulae. For instance, a compound of carbon and hydrogen has a composition corresponding to the formula CH_3. This cannot be made into an adequate structural formula with a quadrivalent carbon atom. But the doubled formula C_2H_6 (for which, of course, there is other evidence as well) leads to the satisfactory formula (3). The number of bonds drawn to each atom must equal its valency.

To maintain the valency rule, it was sometimes necessary to show double or triple bonds, and especially with carbon compounds. For example, ethylene, C_2H_4, and acetylene, C_2H_2, were represented by formulae (4) and (5), rather than by (6) and (7). However,

* The words 'tetravalent', 'divalent' and 'monovalent' are often used, and have become acceptable through usage. We prefer the alternatives given because 'valent' is of Latin origin, whereas 'tetra', 'di' and 'mono' are Greek.

these multiple bonds were by no means an *ad hoc* hypothesis, introduced merely to maintain the quadrivalency of carbon. Ethylene, for instance, has properties of unsaturation consistent with the presence of a double bond in its molecule; ethylene combines with a molecule of hydrogen to yield ethane (8), and this can be reasonably represented as due to the opening up of the double bond:

$$
\begin{array}{c}
\text{H} \\ \diagdown \\
\end{array}
\text{C=C}
\quad + \quad \text{H—H} \quad \longrightarrow \quad
\text{H—C—C—H} \quad (8)
$$

1.5   Stereochemical formulae

The next stage in the evolution of the molecular concept was its extension to the third dimension of space. The formulae we have given so far have necessarily been two-dimensional since they have been written on a sheet of paper. Chemists in 1860 can hardly have supposed that their molecules were really flat objects. Probably they would have felt their formulae were no more than formal symbols, and that—anyway—speculation on the possible deviations of molecules from planarity would be totally unjustified, since there seemed to be no way of testing such ideas. No-one had ever seen a molecule, and it was unlikely that anyone ever would.

Nevertheless, chemistry began to move into the third dimension after 1874, when *stereochemistry* was introduced by le Bel, and—in more detail—by van 't Hoff. The sort of chemical evidence that required this development can be illustrated by a simple example. The hydrogen atoms of the methane molecule (9) can be successively replaced by other atoms, or groups of atoms. If two of them are replaced by (say) chlorine atoms, methylene chloride, CH_2Cl_2, is

$$
\text{H—C—H} \quad (9) \qquad \text{H—C—Cl} \quad (10) \qquad \text{Cl—C—Cl} \quad (11)
$$

produced; and we write two different-looking formulae for it (10) and (11), if we assume—as we surely must—that the incoming chlorine atoms take the places vacated by the outgoing hydrogens. In (10) the chlorines occupy neighbouring sites; in (11) opposite ones. But only one compound of formula CH_2Cl_2 has ever been prepared. Hence either only one of the substances represented by formulae (10) and (11) can exist, or these formulae are really the same formula when stereochemically considered. The latter explanation would be the true one if the four valencies of the carbon atom were directed in space towards the four corners of a tetrahedron. This figure is sketched in (12); it seems intuitively reasonable in that it represents the most 'comfortable' way of arranging four bodies around a central body, if 'comfort' is taken to mean that the four are to be as far away from each other as possible, consistent with their being close to the centre. Corresponding representations of molecules (9), (10) and (11) are drawn in (13), (14) and (15).

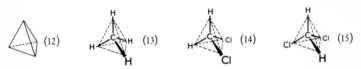

Even without a model, the reader should be able to convince himself that re-orientation of (15) will make it look identical to (14). Such formulae, in which we try to suggest the third dimension by a two-dimensional drawing, are known as *stereochemical formulae*.

Considerations of this kind (often very sophisticated), based on elaborate chemical evidence, led to a detailed knowledge of the three-dimensional relationships between the different parts of molecules. This development was particularly successful in organic chemistry. This is because the valency properties of the carbon atom are particularly simple: given double or triple bonds when necessary, we can always represent the atom as quadrivalent; and the angle between any pairs of the four bonds is rarely far from 110°—the 'tetrahedral angle'.* In inorganic chemistry valency relationships are more complex, but stereochemical knowledge began to grow in that field also, notably from the work of Werner (1866–1919).

As with ordinary structural formulae, nearly all the discoveries of classical stereochemistry, based, as they were, on indirect, chemical evidence, have now been confirmed and amplified by more direct physical methods. Though a metaphysician would, no doubt, question the final validity of the concepts which lie behind any scientific theory, and rightly so, stereochemical theory is now so well-established that a chemist, at any rate, finds it hard to doubt the 'reality' of molecules. Something—he feels—is really 'there' which corresponds rather closely to his stereochemical molecular formulae.

* This angle is more exactly 109° 28′, which is $\cos^{-1}(-1/3)$. This condition arises because, in a regularly tetrahedral molecule, the resolved parts of any three bonds exactly oppose the fourth bond.

1.6   Plan of the book

The general concept of the molecule being taken for granted, the purpose of this small book is to examine some aspects of that concept in more detail. First, we are to consider the forces which bind the atoms together within the molecule. We now have some knowledge of the structures of atoms; they consist of a central nucleus surrounded by electrons. Interactions between atoms to form molecules are largely controlled by the outermost of these electrons. An interpretation of valency in electronic terms is therefore a reasonable first objective. A rigorous treatment is difficult; but a treatment superficial enough to look simple can be helpful as well as useful, provided we do not forget its limitations.

Secondly, we need to consider our knowledge, now very extensive, about the actual sizes, shapes, and other properties of molecules. We shall try to relate this detailed knowledge of molecular properties to our notions of the forces responsible for molecule building. These two approaches supplement each other satisfactorily: the electronic theory of molecules helps us towards a better understanding of molecular structure, whilst experimental study of structure and reactions must always control the development of theory.

1.7   Non-molecular systems

By no means all chemical substances consist of molecules. We must emphasize this limitation on the programme outlined above. A molecule of gaseous hydrogen chloride, HCl, was represented in classical valency theory as H—Cl. Sodium chloride has a composition represented similarly by NaCl, and it was natural to interpret this by a valency-diagram Na—Cl. This suggests that salt consists of Na—Cl molecules, a supposition held by chemists for many years. However, the more powerful methods we now possess have failed to detect any such molecules. We have come to recognise that a salt such as sodium chloride is made up of charged atoms, or *ions*, Na^+ and Cl^-. The oppositely charged ions attract one another strongly; in the sense of the word valency that implies a force of attraction, there is certainly a valency attraction between them. However, in the other sense—of a number of bonds—the situation is different. The ions arrange themselves so as to minimize their total free energy; and just how this may be achieved depends upon the conditions. In the crystalline state, each Na^+ has, as its nearest neighbours, six equidistant Cl^- ions; and conversely. We cannot regard any particular pair as a sodium chloride molecule. In solution in water, the ions are largely independent of one another. Only at very high temperatures, in the vapour, does sodium chloride exist as distinct ion-pairs, which can then be regarded as molecules.

Ions may consist of several atoms. They are bound together by definite bonds, as in a normal molecule, but the whole entity carries a net electrical charge. Simple examples are the sulphate and the ammonium ions: SO_4^{2-} and NH_4^+. They can be regarded as charged molecules.

A different problem is presented by substances like diamond, or graphite, or silica. Diamond, for instance, consists of a crystalline array of carbon atoms, each one strongly bonded to four others situated in tetrahedral directions. Such a crystal consists of atoms rather than of molecules, unless indeed we look upon the whole crystal as a single 'giant molecule', whose 'molecular weight' depends on the weight of the piece.

The book is mainly concerned with valency within the molecule—normally a neutral molecule, but sometimes a charged ion. We must never forget that many important substances cannot be described in molecular terms. Valency theory must try to include such substances also.

1.8   The Avogadro number

One good reason why the early chemists were sceptical about the reality of molecules was that they had no idea—or at most only a very rough idea—of the size of a molecule, or of the number of them in a given quantity of a compound. We now have fairly exact knowledge on this point. The key is the *Avogadro Number*, which is the actual number of molecules (or atoms) in a mole (or g. atom) of any substance. The currently accepted value for this number if $6·0225 \times 10^{23}$ mol./mole, and it is known with an accuracy of about one part in 50,000. The reciprocal of the Avogadro Number is roughly $1·66 \times 10^{-24}$; and if we multiply the molecular weight of any substance by this factor, we obtain the actual weight of the individual molecule. The water molecule thus weighs $18 \times 1·66 \times 10^{-24} = 2·99 \times 10^{-23}$ g. Since the density of liquid water is about $1·0$, it follows that the volume occupied by each molecule has the same value, when expressed as cubic centimetres. If, for simplicity, we suppose this small volume to be a cube, its edge would be $\sqrt[3]{2·99 \times 10^{-23}}$ cm $= 3·1 \times 10^{-8}$ cm. This is a reasonable upper limit for the 'size' of a water molecule; appropriately, it is somewhat larger than the interatomic distances given in §1.2.

1.9 Some natural constants

In addition to the Avogadro Number, the following constants will be needed later:

Planck's constant, $h = 6.626 \times 10^{-27}$ erg-sec;
Electronic charge, $e = 4.803 \times 10^{-10}$ e.s. unit;
Electronic mass, $m = 9.109 \times 10^{-28}$ g;
Velocity of light, $c = 2.9979 \times 10^{10}$ cm/sec;
1 cal $= 4.184$ $J = 4.184 \times 10^7$ erg.

The following table gives some useful conversion factors:

erg/atom or molecule	eV	kcal/mole
1	6.242×10^{11}	1.440×10^{13}
1.602×10^{-12}	1	2.306×10^{1}
6.946×10^{-14}	4.336×10^{-2}	1

Chapter 2

The Atom

2.1 The atom

Before we can describe how atoms combine to yield molecules, we must consider the structure of the atom. Current theory is complicated, but a simple view will suffice here. We may start by thinking of an atom as a spherical body of diameter around 10^{-8} cm. Nearly all its mass is concentrated in a nucleus which is some 10,000 times smaller in diameter. On a volume basis, the nucleus occupies only one part in 10^{12} of the atom, though it carries 99.95 % of the mass. The nucleus also carries a positive charge. The rest of the space is occupied by negatively charged electrons moving in the intense electrical field due to the nucleus.

Electricity, like matter, is atomic; the natural unit of charge is the amount carried by the electron, represented by **e** and equal to about 4.8×10^{-10} electrostatic unit. In terms of this unit, the positive charge carried by the atomic nucleus is Z, the atomic number; this is the integer falling to each when the elements are placed in the natural order suggested by their properties and then numbered off, starting with hydrogen as 1. These numbers are included in Table 2.2 (§ **2.9**). As the atom is normally neutral, the Z positive charges on its nucleus are balanced by Z electrons in the outer spaces of the atom.

It is these electrons which largely determine the properties of the element. Where many electrons surround the nucleus, it is the outermost and loosely held electrons that are most important. The object in this book is to give some idea of the arrangement of these extra-nuclear electrons and of the connexion between this arrangement and chemical and physical behaviour.

What we have just given is an outline sketch of the atom. Many areas within the outline have been deliberately left blank; and, should the reader pursue the study of chemistry further, he will have to fill in these areas with other properties of the atom. For example, the nucleus may possess a 'spin'; if it does, the nucleus will have a magnetic moment and may have an electrical quadrupole moment. Such nuclear properties are now of great importance to the chemist.

However, in the present context, the simplification we have made will do little harm. For many purposes a naive picture of the atom serves the chemist surprisingly well.

2.2   The hydrogen atom

The hydrogen atom, with its single electron, is the simplest atom, and therefore a natural starting point.

Valency properties of atoms (even H) are complicated when we try to account for them quantitatively, as we shall see in Chapter 5. We need to start with something easier if we seek to develop a precise theory of the atom. The most suitable property is the spectrum. As has been known for a long time, the spectrum of atomic hydrogen consists of several series of sharp lines. The most famous is the Balmer series, named after the physicist who first recognised that the wavelengths (λ) of these lines could be accurately represented by a remarkably simple formula. Spectroscopists now prefer to use wave numbers (σ), the connexion being $\sigma = 1/\lambda$; and the wave numbers of the lines in this best-known part of the hydrogen spectrum are given by

$$\sigma = R_H(1/n_1^2 - 1/n_2^2), \qquad (2.1)$$

where $n_1 = 2$ and n_2 may have the values $3, 4, 5 \ldots$, and R_H has the numerical value $109{,}677 \text{ cm}^{-1}$ and is known as the *Rydberg constant*. Other series are also known with $n_1 = 1, 3$ and 4, with the appropriate ranges of values possible for n_2.

The reason for a difference formula of this sort became clearer with the development of the Quantum Theory by Planck (1902). It implies that the energy of the atom is *quantized*: that the amount of energy in the atom is restricted to a number of permissible *energy levels*. ('The energy must run up, or down, the stairs; it cannot slide down the bannisters.') So long as the amount of energy in the atoms stays constant, we do not notice the atom. We do notice it—in the spectrum—when the energy level changes; for then the energy difference, ΔE, is emitted or absorbed as light with a frequency (v) or a wave number (σ) given by the equations,

$$\Delta E = \mathbf{h}v = \mathbf{h}\mathbf{c}\sigma,$$

where $\mathbf{h}$ is Planck's constant and $\mathbf{c}$ is the velocity of light.

We may now redefine our problem: we need to find some theoretical model of the hydrogen atom which will enable us to calculate its energy-levels and hence to account for the spectrum as expressed in the Balmer-type formulae. Figure 2.1 represents some of the possible energy-levels by horizontal lines. Possible transitions, with emission or absorption of light, are indicated by the vertical lines.

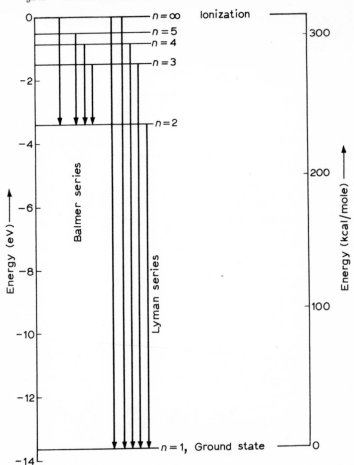

Fig. 2.1 Some energy levels of the hydrogen atom.

The first attempt was made by Bohr in 1913, and it was extremely successful up to a point. Following Rutherford's theory of the nuclear atom, Bohr supposed that the electron would revolve round the nucleus in a circular orbit. He postulated that only certain orbits were allowed, each corresponding to one of the permissible energy-levels shown in the Figure. The greater the radius of the orbit, the higher was the energy. Light was emitted when the electron jumped inwards from a higher to a lower orbit; it was absorbed when the jump was outwards. The permissible orbits were designated by a *quantum number, n*, which had the value 1 for the smallest orbit nearest to the nucleus, and values 2, 3 ... for larger and more distant orbits, until $n = \infty$ corresponded to the complete removal of the electron from the atom. We shall not follow Bohr's treatment here; it is given in many textbooks.* With the help of certain assumptions he was able to calculate values for the energy-levels, and these accounted for the observed spectrum with great accuracy. In his calculation Bohr had to use the accepted values of such constants as **h** and **e**; and, within the limits then imposed by uncertainties affecting these values, his value for the Rydberg constant was indistinguishable from that obtained experimentally.

Despite its success here, and, when elaborated by Sommerfeld, in other directions also, this 'planetary theory' of the H-atom proved to be unacceptable. It contravenes what later came to be recognized as a fundamental principle.

2.3   Wave properties of matter and the uncertainty principle

Light possesses many properties that are consistent with its being a wave-phenomenon; and a wave theory of light had been generally accepted since the early nineteenth century. The Quantum Theory (1902) implied that some other properties of light were after all corpuscular, as Newton had supposed in the seventeenth century. We have the apparent contradiction that some properties can be interpreted in terms of waves and others in terms of particles (quanta). With matter a contrary evolution occurred. The atomic theory, which is so well established, requires us to think of matter as made up of particles—molecules, atoms, nuclei and electrons, for instance. But in 1923 de Broglie suggested that a particle of matter of mass, m, moving with a velocity, v, would be associated with a wave of wavelength $\lambda = \mathbf{h}/(mv)$, where **h** is again Planck's constant. This was soon confirmed by experiment: a beam of electrons was proved to be diffracted by a crystalline material in accordance with the above equation.

Any sort of moving particle will have similar wave-like properties. Neutrons can be diffracted, for instance; and indeed neutron diffraction is now used as a standard method for locating H-atoms in a crystal (see §4.5). However, for heavier particles the waves tend to be shorter and more difficult to detect. It is therefore with the wave-aspect of the lightest particles—electrons—that we are concerned here.

* For example: W. J. Moore, *Physical Chemistry*, 4th edn., Longmans Green, London, 1963, p. 474; A. J. Mee, *Physical Chemistry*, 6th edn., Heinemann Educ., London, 1966, p. 111.

The necessity of supplementing our corpuscular view of the electron with a wave view springs basically from the *Uncertainty Principle*: it is impossible to measure—or even to define—the exact position of a particle and its exact momentum at the same time. If the uncertainty of position is denoted by Δx and that of the momentum, $p \, (= mv)$, by Δp, then

$$\Delta x \times \Delta p \geqslant \mathbf{h}.$$

So when we detect any sort of particle, be it an electron or a photon (quantum) of light, we shall always find the whole unit of matter or of light-energy; we shall never find half an electron or a quarter of a quantum of light-energy. But we cannot be certain of detecting it; the 'when' should be replaced by 'if'. Because of this uncertainty, we have to think in terms of a *probability* of detection; and it is this probability that is governed by wave-properties. More specifically, for an electron there is a *wave-function*, ψ, whose value at different times (t) or in different positions (x) follows a wave-equation of the type, $\psi = A \cos(xt)$; ψ will have alternate positive and negative values in accordance with its periodic, cosinusoidal nature. The chance of finding the electron at point x at time t is proportional to the square of ψ. While ψ can be positive, negative or zero, ψ^2 must always be positive or zero.

Electrons in an atom or molecule are forced to remain close to their appropriate atomic nucleus by the strongly attractive electrostatic field. The electron-waves, ψ, therefore form stationary, or standing, waves, just as do waves of any more familiar sort when they are confined—for instance, the vibrations of a violin string, or the vibrating column of air in a wind instrument. The problem of electrons in atomic systems therefore resolves itself into a study of the possible systems of stationary wave-functions near to an atomic nucleus.

2.4　The problem of the particle in a one-dimensional box

This exercise enables us to gain a simple idea of how energy-levels and quantum numbers arise in a natural manner when we have an electron constrained to abide in a restricted space. The problem is artificial, though it closely parallels the situation in a hydrogen atom. And the mathematics is much simpler.

In Fig. 2.2 we represent a wave function, ψ, which extends in one-dimensional space in the direction x. The wave is confined between the points A and B. When it reaches either limit it is reflected back. A standing wave will then be established provided that the wavelength adjusts itself so that the distance a from A to B is equal to a

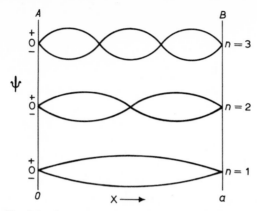

Fig. 2.2   Stationary waves in a one-dimensional box.

whole number (n) of half-wavelengths ($\lambda/2$), i.e.

$$a = n(\lambda/2).$$

This equation holds for any type of waves. When the waves correspond to a particle of mass m, de Broglie's equation can be applied, so that

$$a = n(h/2mv);$$

and, since the (kinetic) energy of a moving mass is given by the relationship $E = \frac{1}{2}mv^2$, we easily reach the result:

$$E = (h^2/8ma^2)n^2.$$

This equation is informative. First, for a given size of box, the possible energy-levels of the system depend only on the value of n, which can be 1, 2, 3 ... etc., corresponding to the various harmonics of the stationary waves, the first three of which are indicated in Fig. 2.2. So n is a quantum number, and it arises because of the wave-nature of the particle. Secondly, the relationship between E and n depends on the factor in parentheses, and this becomes larger the smaller are m and a. Only for a light particle in a small box does the spacing of successive energy-levels become large enough to be important.

For a similar particle in a three-dimensional box with sides a, b and c an analogous equation holds for the energy; but now there are three quantum numbers to be specified before the energy-level becomes defined.

2.5   The wave–mechanical treatment of the hydrogen atom

The theory of the H-atom follows similar lines. We have a three-dimensional problem, and the electron is controlled by the attraction of the positive nucleus rather than by the walls of a box. The waves do not end abruptly at a boundary wall; they die away gradually to a nodal surface, where $\psi = 0$, at infinity. But ψ becomes very small at a certain distance from the nucleus, and we can still regard the atom as having an effectively limited radius. We also have to take into account potential, as well as kinetic, energy. Though mathematically a little more complicated, the outcome is similar. Three quantum numbers are needed to specify the harmonic and hence the energy-level. They are usually represented by the letters n, l and m, and we will explain them shortly. However, for an undisturbed H-atom, the energy is dependent only on the value of n and is given by the equation,

$$E = - \frac{2\pi^2 \mathbf{e}^4 \mathbf{m}}{\mathbf{h}^2} \left(\frac{1}{n^2} \right), \tag{2.2}$$

where $\mathbf{e}$ is the electronic charge and $\mathbf{m}$ its mass, and where n may now be called the *principal quantum number* with $1, 2, 3, \ldots$ etc. as its permissible values. The higher is n, the higher is E. The negative sign is formal only: it means that we have arbitrarily chosen our state of zero energy to be that, with $n = \infty$, in which the electron has been completely removed from the atom.

We see that equation (2.2) leads to a result in accord with the Balmer-type formula (equation 2.1). The energy change, ΔE, when the stationary wave-system changes from that specified by n_1 to that specified by n_2 corresponds to radiation of wave-number σ, where

$$\sigma = \frac{\Delta E}{\mathbf{h} c} = \frac{2\pi^2 \mathbf{e}^4 \mathbf{m}}{\mathbf{h}^3 \mathbf{c}} \left(\frac{1}{n_1^2} - \frac{1}{n_2^2} \right).$$

This is exactly the formula obtained by Bohr; and, when values for the various constants are substituted, the first factor on the right-hand side works out in agreement with the observed value of the Rydberg constant.

Though equation (2.2) suffices for a very close approximation to the energy of an isolated and undisturbed H-atom, it is not generally adequate. Each value of n covers a series of different energy-levels, whose presence is disclosed when the atom is placed in an electrical or magnetic field; and to cover these possibilities two other quantum numbers prove to be necessary. From another point of view, three numbers are needed to specify the particular harmonic adopted by the three-dimensional stationary-wave system which represents the electron when it is held by the attractive force of the nucleus.

The rules for these quantum numbers arise naturally from the mathematical conditions needed to give solutions of the wave-mechanical equation appropriate to this problem. It is beyond the scope of this book to give the mathematical treatment, which is covered in a number of textbooks*. The rules themselves are simple and they are basic for the interpretation of chemistry.

We have already stated that the *principal quantum number*, n, may have any integral value from 1 upwards. The *azimuthal quantum number*, l, may be zero or have any positive, integral value less than n. For instance, when n is 3, l may be 0, 1 or 2. The *magnetic quantum number*, m, may have positive or negative, integral values from $+l$ to $-l$, including zero. So, when l is 2, m may have any of the five values, -2, -1, 0, $+1$ or $+2$. In general, there are $(2l + 1)$ possible values for m.

2.6 Orbitals

A permitted set of values for these three quantum numbers, such as $n = 3$, $l = 1$ and $m = -1$, serves to define a possible electronic situation in the H-atom. Any such possibility is known as an *orbital*, a term used because an orbital in wave mechanics corresponds to a possible electronic orbit in the planetary model of the atom.

A possible orbit for a planet moving round the sun is a mathematical fiction until a body actually moves in that path. In the same way, the orbitals of an H-atom are possible ways in which the electron-waves may vibrate. As an isolated H-atom has only one electron, only a single orbital is occupied at any time. We observe the atom spectros-copically when the electron has moved from one orbital to some other.

The possible orbitals are restricted to the sets of quantum numbers allowed by the rules given above. They can conveniently be set out in the form of Table 2.1, which however could be extended indefinitely towards the right. The reader should check that when $n = 4$ there are sixteen possible orbitals; and that in general there are n^2 possibilities.

Orbitals like these which are associated only with one single atom are known as *atomic orbitals* (a.o.). A group of orbitals with a common value of n are sometimes known as a *shell* (i.e. of electrons); and the capital letters K, L, M, N, etc. have become associated with the shells

* C. A. Coulson, *Valence*, 2nd edn., Oxford University Press, London, 1961; C. W. N. Cumper, *Wave Mechanics for Chemists*, Heinemann Educ., London, 1966; J. W. Linnett, *Wave Mechanics and Valency*, Methuen, London; Wiley, New York, 1960.

Table 2.1

n	1	2			3		
l	0	0	1		0	1	2
m	0	0	−1 0 +1		0	−1 0 +1	−2 −1 0 +1 +2
Number of orbitals in sub-group	1	1	3		1	3	5
Designation	1s	2s	2p		3s	3p	3d
Total number of orbitals in group or shell	1	4			9		
Designation of shell	K	L			M		

having $n = 1, 2, 3, 4$, etc. The orbitals of a sub-group with a common value of l are denoted by a lower-case letter according to the following code: s for $l = 0$, p for 1, d for 2, f for 3. (The first four letters stand for 'sharp', 'principal', 'diffuse' and 'fundamental', words applied by spectroscopists to certain types of spectra at a time when nothing was known about atomic structure. Subsequent letters follow alphabetically.)

The azimuthal quantum number, l, is a measure of the *angular momentum* of the atom in so far as this is due to the orbital motion of the electron.

2.7   The shapes of atomic orbitals

The wave-mechanics of the H-atom can be solved rigorously. Solutions of the equation occur when appropriate values are allocated to the three quantum numbers, and from these solutions we can deduce the shapes of the various orbitals. We can find the shape of the stationary ψ-wave system representing the distribution of the chance of finding the electron when it occupies that particular orbital. These shapes turn out to be of great value for our understanding of the properties not only of the H-atom, but also of those of other elements.

An s-orbital has the simplest shape. It is spherically symmetrical; it dies away equally rapidly in all directions. Diagrammatically we can represent an s-orbital in several ways. Figure 2.3(a) shows how ψ diminishes with distance (r, in any, and every, direction) from the central nucleus; ψ vibrates between positive and negative values, and what we have plotted is the decreasing amplitude of the stationary wave as r increases. Of more direct interest is ψ^2, which cannot be negative, and which represents the probability of finding the electron at a given point. This, which we show in Fig. 2.3(b), has the same sort of shape as ψ but falls off more rapidly; ψ^2 becomes zero only when $r = \infty$, though it becomes small when r is greater than 1 Å and negligible for most purposes beyond 2 Å. We can suggest the physical meaning of this by Fig. 2.3(c) which is meant to convey an impression of the electron density fading away with distance. We can imply the same notion by the much simpler diagram at (d), where the circle symbolizes a sphere within which there is a (specified) high probability of finding the 1s-electron.

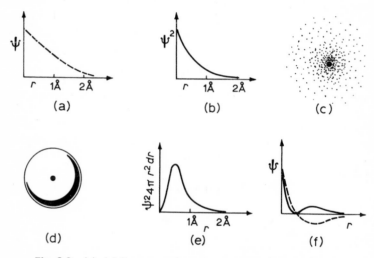

Fig. 2.3   (a)–(e) Representations of a 1s-orbital; (f) a 2s-orbital.

Now we come to Fig. 2.3(e), giving the *radial-distribution function*, $\psi^2 4\pi r^2 \, dr$, which measures the probability of finding the electron at a given distance r. At first sight, the reader might suppose that the maximum probability would be at the centre (where $r = 0$) since ψ and ψ^2 are themselves at a maximum there. However, the centre is a mathematical point, of zero volume, so that the chance of finding the electron there is zero; as we move outwards the amount of space, at a given distance, increases in proportion to the area ($4\pi r^2$) of a sphere of radius r; and at first this increase more than compensates for the diminution of ψ^2. This we arrive at a curve of the shape shown. The maximum occurs at $r = 0.53$ Å, which is known as the *Bohr radius*, because this was the radius found by Bohr for his first orbit, of lowest energy, nearest to the nucleus. There is this *correspondence* between the semi-classical theory of Bohr and the findings of wave mechanics.

All *s*-orbitals are spherically symmetrical; but, when n is greater than 1, they become more complicated. For example, ψ for a 2*s*-electron has the shape shown by the broken line in Fig. 2.3(f): it starts positive (or negative), passes through zero, and then becomes negative (or positive) as it dies away outwards; ψ^2, which cannot be negative, is represented by the continuous line. This means that the 2*s*-orbital has a spherical node, where both ψ and ψ^2 are zero; the electron may be either inside or outside this surface; the electron-density cloud has two 'skins like an onion'.

Sometimes one is asked the question how the electron can get from the inner to the outer skin when, to do so, it must apparently cross the nodal sphere where ψ^2 is zero. The question is not well based, as we can see by looking at Fig. 2.2 and supposing that the case with $n = 2$ now represents a string vibrating in its second harmonic. The central node is a point where the string has zero lateral vibration; nevertheless the vibrations on either side of it are connected, and together constitute the total mode of vibration of the string.

There are three equivalent *p*-orbitals for a given value of n. Each has a nodal plane passing through the nucleus, and each orbital extends perpendicularly to this plane. The three orbitals are mutually at right angles, and we may suppose them to be directed along x-, y- and z-axes of coordinates. Figure 2.4 represents the 2p_x-orbital in

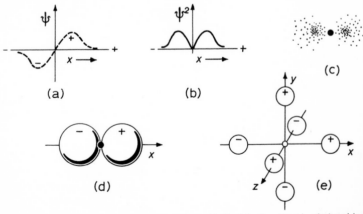

Fig. 2.4 (a)–(d) Representation of a 2*p*-orbital; (e) directional relationships of the three 2*p*-orbitals.

various conventions: (a) shows ψ as having a positive amplitude to the right and negative to the left. This means that the phases of ψ are opposed; when one side is 'up' the other is 'down', ψ being always zero in the yz-plane where $x = 0$. The electron density, being proportional to ψ^2, has the form shown in Fig. 2.4(b), and this is also represented in (c) and (d) which follow the conventions explained above for Fig. 2.3(c) and (d). There is no need to draw the p_y- and p_z-orbitals; they are identical in shape, but directed along the other two orthogonal axes, as suggested by Fig. 2.4(e).

The $+$ and $-$ signs for ψ marked on the lobes of the p-orbitals are formal in the sense that the electron density, measured by ψ^2, must necessarily be positive. They are significant, however, as we shall see later, when we consider how different orbitals may be combined. (There is admittedly room for initial misunderstanding when ψ may be either positive or negative, when we explain that ψ^2 must always be positive, and when we add that this positive ψ^2 measures the density of negative charge . . .)

When n exceeds 2, the p-orbitals have additional nodal surfaces, but the directional property of the three orbitals still applies as sketched in Fig. 2.4(e). When the three p-orbitals are suitably combined together, they add up to a totality which possesses spherical symmetry (see §2.12).

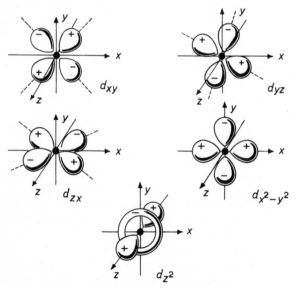

Fig. 2.5   The five (equivalent) 3d-orbitals.

The shapes of the five d-orbitals, with a given value of n, are more complex. These shapes are represented in the conventional manner by Fig. 2.5. Each orbital has two nodal surfaces: mutually perpendicular planes in the first four, and two conical surfaces in the last. As with the three p-orbitals, the five d-orbitals can be combined to yield spherical symmetry. The five orbitals are also equivalent. The reader will easily agree with this statement as it affects the first four d-orbitals; but he may be perplexed by the odd appearence of the d_z^2-orbital. This arises out of the problem of dividing a spherical totality, according to certain rules, into five parts.

2.8   Atoms with more than one electron

As the number of electrons in an atom increases beyond one, a rigorous wave-mechanical treatment becomes difficult and soon impossible. Nevertheless an understanding of these more complex atoms can be based on the orbitals found in the H-atom. The orbitals are changed: for one thing, the electron density tends to be drawn towards the centre by the additional nuclear charge; for another, the presence of extra electrons affects the energy-levels in a complicated way. But the orbitals are still of the type summarized in Table 2.1, and their shapes correspond to those described in the previous section.

Two principles indicate the arrangement of many electrons in an atom. The first is the *Pauli Principle*, which can be simply stated thus: no two electrons in the same atom may have the same values of all (four) quantum numbers. An a.o. is specified by giving values to the three quantum numbers, n, l and m. A fourth is needed to take account of the phenomenon of *electron spin*. An electron has a magnetic moment; the term 'spin' is associated with this because the moment would arise if it were a rotating, negatively-charged sphere, though this may not be literally true. According to quantum principles, this magnetic moment may be orientated in two ways—with, or against, an external magnetic field. These possibilities are covered by introducing the fourth, *spin-quantum number*, s, which is restricted to the values $-\frac{1}{2}$ and $+\frac{1}{2}$: two values that are equal in magnitude, opposite in sign, and differing by unity. The Pauli Principle includes spin, so that, when we have a set of values for n, l and m, we can complete the specification by assigning $-\frac{1}{2}$ or $+\frac{1}{2}$ to s. There is no other possibility. Therefore a given orbital cannot contain more than two electrons; if there are two, they must have opposite spins.

The second principle concerns the order in which the appropriate number of electrons is fed into the available orbitals. Normally we are concerned with the atom in its *ground state*—its state of lowest energy. It follows almost as a truism that the electrons will go into the a.o. in the order of their increasing energy. In the unperturbed H-atom the energy of an electron is almost wholly determined by the value of n; to a very close approximation, whether the electron is in $3s$, or $3p$, or $3d$ makes no difference. However, when we are adding the eleventh electron to complete the requirements of a neutral Na-atom, there is a considerable difference: in that sense, $3s < 3p < 3d$. The three types of sub-level are now split by the presence of the inner electrons, and the Na-atom is in its ground state when the final electron is in the $3s$-orbital. The electronic structure of the atom can be represented by $1s^2, 2s^2, 2p^6, 3s^1$. The splitting of sub-levels is illustrated in Fig. 2.6—a diagram which is not to scale, but which places the sub-levels in the correct order: $1s, 2s, 2p, 3s, 3p, 4s, 3d, 4p, 5s, 4d, 4f$, $5p$, etc. This sequence is of fundamental importance.

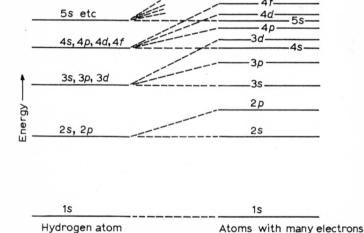

Fig. 2.6   Splitting of sub-levels.

2.9   The structure of the periodic table

The electronic arrangements in the atoms*—each in its ground state—of the elements are set out in Table 2.2, which should be studied in the light of the following commentary.

The first and second electrons, in H and He, are located in the lowest orbital, $1s$. The K-shell is then filled, and He is an inert gas. The third electron in Li must go into the higher L-shell; it is accommodated in the $2s$-orbital, which is filled when it receives the fourth electron

* Strictly, this applies to the gaseous, elementary atom.

Table 2.2 Arrangement of Electrons in Groups and Sub-Groups

Principal Quantum Number, n		1	2		3			4				5			6			7	
Group, or Shell		K	L		M			N				O			P			Q	
Sub-Group		1s	2s	2p	3s	3p	3d	4s	4p	4d	4f	5s	5p	5d	6s	6p	6d	7s	I
Element and Atomic Number																			eV
H	1	1																	13·6
He	2	2																	24·6
Li	3	2	1																5·4
Be	4	2	2																9·3
B	5	2	2	1															8·3
C	6	2	2	2															11·3
N	7	2	2	3															14·6
O	8	2	2	4															13·6
F	9	2	2	5															17·4
Ne	10	2	2	6															21·6
Na	11	2	2	6	1														5·1
Mg	12	2	2	6	2														7·6
Al	13	2	2	6	2	1													6·0
Si	14	2	2	6	2	2													8·2
P	15	2	2	6	2	3													11·0
S	16	2	2	6	2	4													10·4
Cl	17	2	2	6	2	5													13·0
Ar	18	2	2	6	2	6													15·8

Z	Element	1s	2s	2p	3s	3p	3d	4s	4p	4d	5s	
19	K	2	2	6	2	6		1				4·3
20	Ca	2	2	6	2	6		2				6·1
21	Sc	2	2	6	2	6	1	2				6·7
22	Ti	2	2	6	2	6	2	2				6·8
23	V	2	2	6	2	6	3	2				6·7
24	Cr	2	2	6	2	6	5	1				6·8
25	Mn	2	2	6	2	6	5	2				7·4
26	Fe	2	2	6	2	6	6	2				7·9
27	Co	2	2	6	2	6	7	2				7·8
28	Ni	2	2	6	2	6	8	2				7·6
29	Cu	2	2	6	2	6	10	1				7·7
30	Zn	2	2	6	2	6	10	2				9·4
31	Ga	2	2	6	2	6	10	2	1			6·0
32	Ge	2	2	6	2	6	10	2	2			8·1
33	As	2	2	6	2	6	10	2	3			10·5
34	Se	2	2	6	2	6	10	2	4			9·8
35	Br	2	2	6	2	6	10	2	5			11·8
36	Kr	2	2	6	2	6	10	2	6			14·0
37	Rb	2	2	6	2	6	10	2	6		1	4·2
38	Sr	2	2	6	2	6	10	2	6		2	5·7
39	Y	2	2	6	2	6	10	2	6	1	2	6·7
40	Zr	2	2	6	2	6	10	2	6	2	2	6·9
41	Nb	2	2	6	2	6	10	2	6	4	1	—
42	Mo	2	2	6	2	6	10	2	6	5	1	—
43	Tc	2	2	6	2	6	10	2	6	6	1	—
44	Ru	2	2	6	2	6	10	2	6	7	1	—
45	Rh	2	2	6	2	6	10	2	6	8	1	—
46	Pd	2	2	6	2	6	10	2	6	10		—

Table 2.2 (*contd.*)

Principal Quantum Number, *n*	1	2		3			4				5			6			7	*I*
Group, or Shell	*K*	*L*		*M*			*N*				*O*			*P*			*Q*	eV
Sub-Group	1*s*	2*s*	2*p*	3*s*	3*p*	3*d*	4*s*	4*p*	4*d*	4*f*	5*s*	5*p*	5*d*	6*s*	6*p*	6*d*	7*s*	eV
Element and Atomic Number																		
Ag 47	2	2	6	2	6	10	2	6	10		1							
Cd 48	2	2	6	2	6	10	2	6	10		2							
In 49	2	2	6	2	6	10	2	6	10		2	1						
Sn 50	2	2	6	2	6	10	2	6	10		2	2						
Sb 51	2	2	6	2	6	10	2	6	10		2	3						
Te 52	2	2	6	2	6	10	2	6	10		2	4						
I 53	2	2	6	2	6	10	2	6	10		2	5						
Xe 54	2	2	6	2	6	10	2	6	10		2	6						
Cs 55	2	2	6	2	6	10	2	6	10		2	6		1				
Ba 56	2	2	6	2	6	10	2	6	10		2	6		2				
La 57	2	2	6	2	6	10	2	6	10		2	6	1	2				
Ce 58	2	2	6	2	6	10	2	6	10	1	2	6	1	2				
Pr 59	2	2	6	2	6	10	2	6	10	2	2	6	1	2				
Nd 60	2	2	6	2	6	10	2	6	10	3	2	6	1	2				
Pm 61	2	2	6	2	6	10	2	6	10	4	2	6	1	2				
Sm 62	2	2	6	2	6	10	2	6	10	5	2	6	1	2				
Eu 63	2	2	6	2	6	10	2	6	10	6	2	6	1	2				
Gd 64	2	2	6	2	6	10	2	6	10	7	2	6	1	2				

Tb	65	2	2	6	2	6	10	2	6	10	8	2	6	1	2			
Dy	66	2	2	6	2	6	10	2	6	10	9	2	6	1	2			
Ho	67	2	2	6	2	6	10	2	6	10	10	2	6	1	2			
Er	68	2	2	6	2	6	10	2	6	10	11	2	6	1	2			
Tm	69	2	2	6	2	6	10	2	6	10	12	2	6	1	2			
Yb	70	2	2	6	2	6	10	2	6	10	13	2	6	1	2			
Lu	71	2	2	6	2	6	10	2	6	10	14	2	6	1	2			
Hf	72	2	2	6	2	6	10	2	6	10	14	2	6	2	2			
Ta	73	2	2	6	2	6	10	2	6	10	14	2	6	3	2			
W	74	2	2	6	2	6	10	2	6	10	14	2	6	4	2			
Re	75	2	2	6	2	6	10	2	6	10	14	2	6	5	2			
Os	76	2	2	6	2	6	10	2	6	10	14	2	6	6	2			
Ir	77	2	2	6	2	6	10	2	6	10	14	2	6	7	2			
Pt	78	2	2	6	2	6	10	2	6	10	14	2	6	8	2			
Au	79	2	2	6	2	6	10	2	6	10	14	2	6	10	1			
Hg	80	2	2	6	2	6	10	2	6	10	14	2	6	10	2			
Tl	81	2	2	6	2	6	10	2	6	10	14	2	6	10	2	1		
Pb	82	2	2	6	2	6	10	2	6	10	14	2	6	10	2	2		
Bi	83	2	2	6	2	6	10	2	6	10	14	2	6	10	2	3		
Po	84	2	2	6	2	6	10	2	6	10	14	2	6	10	2	4		
At	85	2	2	6	2	6	10	2	6	10	14	2	6	10	2	5		
Rn	86	2	2	6	2	6	10	2	6	10	14	2	6	10	2	6		
Fr	87	2	2	6	2	6	10	2	6	10	14	2	6	10	2	6		1
Ra	88	2	2	6	2	6	10	2	6	10	14	2	6	10	2	6		2
Ac	89	2	2	6	2	6	10	2	6	10	14	2	6	10	2	6	1	2
Th	90	2	2	6	2	6	10	2	6	10	14	2	6	10	2	6	2	2
Pa	91	2	2	6	2	6	10	2	6	10	14	2	6	10	2	6 ?	3	2 ?
U	92	2	2	6	2	6	10	2	6	10	14	2	6	10	2	6	4	2

in Be. In the next six elements, B to Ne, six electrons go successively into the three $2p$-orbitals. At Ne the L-shell is complete, and Ne is a second inert gas, with an 'octet' of electrons in its outer shell ($2s^2$ and $2p^6$).

The special stability of the octet in Ne is not surprising since it corresponds to the filling of all four orbitals in the second shell. It is more surprising to find that an octet confers similar stability in higher shells also, even though it does not then comprise all the orbitals of the shell. Each of the inert gases has a completed octet as its outermost group.

The elements from Li to Ne constitute the *First Short Period**. The initiation and completion of an octet in the $3s$- and $3p$-orbitals gives a *Second Short Period*, Na to Ar; Ar is an inert gas, despite the vacant state of its $3d$-orbitals. The *First Long Period*, comprising the eighteen elements from K to Kr, involves the completion of an octet in the $4s$- and $4p$-orbitals. A new feature is the interruption of this process whilst the vacancies in the five $3d$-orbitals are filled. This gives a sequence of ten *transitional elements*, Sc to Zn, between the first two and the last six elements which make up the strict New-landsian octave. The interpretation is given by the order of the sub-levels in Fig. 2.6: $3d$ comes between $4s$ and $4p$. A *Second Long Period* follows the same pattern: the octet is initiated in Rb and Sr; is interrupted while the $4d$-orbitals are filled in the *second transitional elements*, Y to Cd; and is completed with the six elements In to Xe.

The next period—usually known as the Third Long Period, though we may prefer to call it the *First Very Long Period*—is more com-plicated because it suffers a double interruption. The octet begins to be filled with Cs and Ba ($6s$); the process is interrupted by the filling of the $5d$-orbitals in the *third* set of *transitional elements*, La to Hg; and this interruption is itself held up by the filling of the $4f$-orbitals in the fourteen rare-earth, or *lanthanide*, *elements*, Ce to Lu; whilst finally the octet is completed with Tl to Rn ($6p$). The *Second Very Long Period* is incomplete, despite the discovery of artificial elements beyond U. It probably does not exactly follow the pattern of the preceding period. The sketch of the Periodic Table given in this section is em-bodied in Table 2.3.

2.10 Optical and X-ray spectra and the Hund Rule

The spectra of atoms are usually found in the visible and ultra-violet regions. With wavelengths probably in the range 1000–5000 Å,

---

*Some inorganic chemists regard H and He as constituting the First Period, in which case Li–He constitutes the Second Period, and so on.

Table 2.3 Periodic Table

Group	O	I or Ia	II or IIa	IIIa	IVa	Va	VIa	VIIa	← VIII →	Ib	IIb	III or IIIb	IV or IVb	V or Vb	VI or VIb	VII or VIIb	O	Electronic Sub-Groups being Filled
		H 1															He 2	2 Elements (1s)
First Short Period	He 2	Li 3	Be 4									B 5	C 6	N 7	O 8	F 9	Ne 10	8 Elements (2s, 2p)
Second Short Period	Ne 10	Na 11	Mg 12									Al 13	Si 14	P 15	S 16	Cl 17	A 18	8 Elements (3s, 3p)
First Long Period	Ar 18	K 19	Ca 20	Sc 21	Ti 22	V 23	Cr 24	Mn 25	Fe 26 Co 27 Ni 28	Cu 29	Zn 30	Ga 31	Ge 32	As 33	Se 34	Br 35	Kr 36	18 Elements (4s, 3d, 4p)
Second Long Period	Kr 36	Rb 37	Sr 38	Y 39	Zr 40	Nb 41	Mo 42	Tc 43	Ru 44 Rh 45 Pd 46	Ag 47	Cd 48	In 49	Sn 50	Sb 51	Te 52	I 53	Xe 54	18 Elements (5s, 4d, 5p)
First Very Long Period	Xe 54	Cs 55	Ba 56	La 57 Lanthanides 58–71	Hf 72	Ta 73	W 74	Re 75	Os 76 Ir 77 Pt 78	Au 79	Hg 80	Tl 81	Pb 82	Bi 83	Po 84	At 85	Rn 86	32 Elements (6s, 5d, 4f, 6p)
Second Very Long Period	Rn 86	Fr 87	Ra 88	Ac 89 Actinides 90–?														About 18 Elements known (1965) (7s, ?)

Transitional Elements

they correspond to energy changes around 4–20×10^{-12} erg/atom, or 60–300 kcal/mole. Such changes can be attributed to transitions of the outer electrons from one orbital to a neighbouring orbital. Usually only the electron, or electrons, of the outermost shell are concerned in optical spectra. Details of such spectra are the experimental evidence for the electronic structures shown in Table 2.2.

To disturb the inner electrons requires much greater amounts of energy, corresponding to observable effects at much shorter wavelengths. For instance, the $K\alpha$ X-rays from a Cu-target—the radiation often used in crystal-structure analysis—have a wavelength of about $1\cdot5$ Å, which implies energy changes of about $1\cdot3 \times 10^{-8}$ erg/-atom, or 2×10^{5} kcal/mole. These X-rays are produced by bombarding the Cu-target with a beam of electrons sufficiently energetic to dislodge one of the K-shell ($1s$) electrons of the Cu-atom. The vacancy thus produced is soon filled by an electron from the L-shell ($2s$ or $2p$), and this transition causes emission of the $K\alpha$-radiation.

Spectroscopic evidence leads to the *Hund Rule of Maximum Multiplicity*, which applies to the situation where a sub-group of orbitals is not yet completely filled with electrons. For example, in the isolated N-atom there are only three electrons in the three $2p$-orbitals. If we represent these orbitals by three 'boxes' and the electron with its spin by an arrow, $\uparrow$, there are two alternative arrangements, of which the following symbolization is probably self-explanatory: $\boxed{\uparrow\downarrow|\uparrow|}$ and $\boxed{\uparrow|\uparrow|\uparrow}$. In simple terms, the rule states that the second arrangement obtains; the electrons spread themselves so as to occupy (partially) as many orbitals as possible. The Hund rule has implications amongst the transitional and lanthanide elements, where there is usually incomplete occupancy of the five d- or seven f-orbitals.

2.11   The ionization energy

This is the amount of energy needed to remove one electron right away from a (gaseous) atom, initially in its ground state, so as to yield a singly charged ion. (Greater amounts of energy are needed to remove a second, or a third, electron; and these are termed the second, or third, ionization energies.) It is customary to record these energies as electron-volts (eV)—a unit which is defined as the energy conferred on an electron when it has been accelerated through a potential-difference of one volt; and they are then represented by the symbol I.

For the H-atom I can easily be calculated from equation (2.2), since complete removal of the electron means that the initial state has $n = 1$ and the final state $n = \infty$. Substitution of values for the various constants in the factor $(2\pi^2 e^4 m/h^2)$, we find

$$I = 2 \times 3\cdot142^2 \times (4\cdot8 \times 10^{-10})^4 \times 9\cdot1 \times 10^{-28} \div (6\cdot6 \times 10^{-27})^2,$$

$$= 2\cdot175 \times 10^{-10} \text{ erg/atom};$$

and since $1\,eV = 1\cdot602 \times 10^{-12}$ erg, $I = 13\cdot58\,eV$ (Alternatively $I = 313$ kcal/mole; see §1.9.) This value is in agreement with the experimental one.

Values of I for a number of elements are included in Table 2.2. Their trend can be correlated with position in the Periodic Table. I is at a maximum for the inert gases, whose atoms have the most stable electronic structures. It falls sharply to the next element, an alkali metal, where there is only a single electron in the outermost shell, and where this electron is removed easily; and it tends to rise steadily towards the next inert gas. Within this rise there are some minor singularities. That at Be, for instance, can be related to the completion of the $2s$ sub-level with this element.

2.12   The shapes of atoms

An atom with a completed octet of electrons in its outermost shell is spherical. This is connected with some consideration put forward at the end of §2.7. It applies not only to the neutral atoms of the inert gases, but also to ions which have attained an octet by gain or loss of electrons, e.g. Cl^-, O^{2-}, or Ca^{2+}. Atoms, or ions, with other than a complete octet for the outermost shell (such as O, or K or Cu^{2+}) will not be exactly spherical. The deviation is not usually very considerable. The inner electrons give a high electron density near to the centre, and this will be spherically symmetrical because the inner shells are usually complete; compared with this, lack of exact symmetry in the relatively small electron density near the periphery of the atom is relatively unnoticeable.

Chapter 3

The Elementary Electronic Interpretation of Valency

3.1 History

As we saw in Chapter 1, the doctrine of valency, both as a number and as a force, was well established in chemistry long before 1900. During the early years of the twentieth century the electronic structure of the atom began to be understood, and it was natural to try to use it to explain valency. After some early speculations of limited applicability, theories of wider utility were suggested by G. N. Lewis and by Kossel in 1916; they were developed by Langmuir in 1922–3; and they were elegantly codified by N. V. Sidgwick in his book, *The Electronic Theory of Valency* (1927).

At this stage the electronic interpretation of valency was purely descriptive. It lacked the quantitative insight that was to come later from the application of wave mechanics. Nevertheless what we may now term the 'classical electronic theory of valency' is still a useful point of departure for more up-to-date and sophisticated enquiry. To be able to draw the *classical electronic bond-diagram* is an essential facility for the chemist or the student of chemistry. In this chapter we present the rules for writing electronic formulae. In Chapter 5 we shall turn to the problems of trying to understand these formulae at a deeper level, and in a more quantitative way.

3.2 The inert gases

At its most elementary level, our theory starts from the fact that the inert gases are monatomic and chemically very unreactive. Though inert-gas compounds, such as XeF_2, have been discovered since 1962, they cannot be ranked with the great majority of stable, ordinary compounds. The theory therefore supposes that the electronic groupings existing in the atoms of these gases are peculiarly stable ones, and that atoms which do not initially possess such a grouping tend to attain one by gain or loss of electrons, whereby valency forces are set up. The arrangement of electrons in the atoms of the inert gases is shown in Table 3.1. These are the most stable groupings, though there are others, as we shall see later.

Table 3.1 Electronic Groupings in the Inert Gases

Shell		K	L	M	N	O	P
Element	Atomic Number						
He	2	2					
Ne	10	2	8				
Ar	18	2	8	8			
Kr	36	2	8	18	8		
Xe	54	2	8	18	18	8	
Rn	86	2	8	18	32	18	8

3.3 Electrovalency (and covalency)

Atoms may attain a more stable electronic grouping in two ways: one atom may give one or more electrons to some other atom, or atoms; two atoms may share one or more pairs of electrons. In the first case ions are produced which are said to be united by *electrovalency*; in the second a *covalent bond* is said to have been established.

Electrovalency is neatly illustrated by the combination of elements which are one place on either side of an inert gas. The Cl-atom has seventeen electrons whose arrangement, shown in Table 2.2, can be written: $1s^2 2s^2 2p^6 3s^2 3p^5$; there are seven electrons in the incomplete M-shell. The K-atom has nineteen electrons: $1s^2 2s^2 2p^6 3s^2 3p^6 4s^1$; there is a complete octet in the M-shell and a single electron in N. Between these elements lies Ar, with the complete octet in its outermost shell characteristic of most inert-gas atoms. By transferring one electron from K to Cl both atoms achieve electronic groupings like that of Ar, but the former becomes a positively charged K^+-ion, and the latter become a negatively charged Cl^--ion.

Following a well-known convention, we can represent this process by the statement,

$$\text{K} \cdot \; + \; \cdot \ddot{\underset{\cdot\cdot}{\text{Cl}}} : \; \longrightarrow \; \text{K}^+ \; + \; : \ddot{\underset{\cdot\cdot}{\text{Cl}}} :^- .$$

Dots have been used to represent electrons; but only those of the outermost (or 'valency') shell have been shown. We are also using the chemical symbols K and Cl in a slightly unorthodox sense; for they here represent not the whole atom, but only the 'core'—the residue remaining when the outermost shell of electrons has been removed. The combination of Mg and O can be similarly represented by

$$\text{Mg} : \; + \; \ddot{\underset{\cdot\cdot}{\text{O}}} : \; \longrightarrow \; \text{Mg}^{2+} \; + \; : \ddot{\underset{\cdot\cdot}{\text{O}}} :^{2-} .$$

This form of chemical union is sometimes said to be *via* 'ionic bonds'. We shall not use this term since we prefer to confine the word 'bond' to unions that are, at least mainly, covalent in character. In K^+Cl^-, or $Mg^{2+}O^{2-}$, there is a general attraction between all positive and negative ions, rather than a one-to-one pairing, as we have already explained in § 1.7.

Formally we might expect the applicability of electrovalency to be wide: Mg could attain an inert-gas structure either by loss of two electrons or by gain of six, to yield Mg^{6-}. The latter does not happen. Electrovalency is restricted to the gain or loss of a small number of electrons only—rarely more than two or three. To withdraw successive electrons from an atom becomes increasingly difficult by reason of the increasing positive charge against which the next electron has to be removed. Polyvalent cations are energetically unfavourable, and simple cations rarely have a valency higher than $+3$. Addition of electrons to yield a polyvalent anion is even more difficult: a high negative charge would enhance the mutual repulsion between the electrons and make the ion unstable. Simple anions of valency more than -2 are probably unknown.

The entities bound by electrovalency are not necessarily single atoms. They may be charged covalent molecules, as we stated in § 1.7. The forces between the ions of ammonium sulphate are electrovalent; within each ion the bonding is covalent.

Though the octet is usually present in the outermost shell of the most stable ions, other groupings are possible, especially with the transitional element. The Cu^{2+}-ion has the electronic structure: $1s^2\, 2s^2\, 2p^6\, 3s^2\, 3p^6\, 3d^9$.

3.4   Covalency

Electrovalency may account for chemical combination between atoms of opposite chemical character—between electropositive elements like Na or Ca, and electronegative ones like F or S. But combination of comparable strength also occurs between elements of like character. Chlorine forms rather stable Cl_2 molecules, whilst the intricacies of organic chemistry depend on the ability of the C-atom to combine with other C-atoms to an almost unlimited extent. An effective electronic theory of valency had to wait until Lewis suggested a means of explaining such union of like atoms; he introduced the notion of *covalency*, though the word was coined by Langmuir a few years later.

Sharing of pairs of electrons is a simple way of enabling two atoms to complete octets which lack only a few electrons. This is, of course, a 'paper transaction'; but it covers the observed facts of valency in a great majority of molecules. Here are representations of the formation of the molecules, Cl_2, CH_3F and H_2CO by this covalent process:

$$:\ddot{C}l\cdot \; + \; \cdot\ddot{C}l: \longrightarrow :\ddot{C}l:\ddot{C}l: \, ,$$

$$3H\cdot \; + \; :\ddot{F}\cdot \; + \; \cdot\dot{C}\cdot \longrightarrow H:\ddot{C}:\ddot{F}: \, ,$$
with H above and H below the C.

and

$$2H\cdot \; + \; \cdot\dot{C}\cdot \; + \; \ddot{O}: \longrightarrow \; C::\ddot{O}: \; .$$
with H above and H below the C.

As before the dot* stands for the electron. The Cl-, F-, C- and O-atoms have all attained octets, provided the electrons of the shared pairs are allowed to count for each atom: the H-atom attains the pair of electrons (duplet) characteristic of He.

When we compare the above electronic formulae with the corresponding classical valency formulae,

$$Cl-Cl, \qquad \begin{array}{c} H \\ | \\ H-C-F, \\ | \\ H \end{array} \qquad \begin{array}{c} H \\ \diagdown \\ C=O, \\ \diagup \\ H \end{array}$$

$$(1) \qquad\qquad (2) \qquad\qquad (3)$$

we notice that a shared pair of electrons always appears as a bond-line in (1)–(3). The classical bond therefore took on a new significance when this was first appreciated. The three original electronic formulae can alternatively be written as follows:

$$:\ddot{C}l-\ddot{C}l: \, , \qquad \begin{array}{c} H \\ | \\ H-C-\ddot{F}: \, , \\ | \\ H \end{array} \qquad \begin{array}{c} H \\ \diagdown \\ C=\ddot{O}: \; . \\ \diagup \\ H \end{array}$$

$$(4) \qquad\qquad (5) \qquad\qquad (6)$$

* We avoid the use of symbols such as $\times$ and $\circ$ to distinguish electrons from different atoms (e.g. $\overset{\times\times}{\underset{\times\times}{\times}}\ddot{C}l\times \; + \; \overset{\circ\circ}{\underset{\circ\circ}{\circ}}\ddot{C}l\circ \longrightarrow \overset{\times\times}{\underset{\times\times}{\times}}\ddot{C}l \circ \overset{\circ\circ}{\underset{\circ\circ}{\ddot{C}l}}\circ$). It is a principle of wave mechanics that electrons are indistinguishable.

Since dots are troublesome to write, we shall in this book prefer another convention. As the line is taken to symbolize a shared pair of electrons in (4)–(6), it seems logical to use it also to represent an unshared pair of electrons, or a *lone pair*, as they are often called. With this added convention, the same three molecules can be written,

$$|\overline{\underline{Cl}}\!-\!\overline{\underline{Cl}}| \qquad H\!-\!\overset{\displaystyle H}{\underset{\displaystyle H}{\overset{|}{\underset{|}{C}}}}\!-\!\overline{\underline{F}}| \qquad \overset{\displaystyle H}{\underset{\displaystyle H}{}}\!\!\diagdown\!\!\diagup C\!=\!O\diagup\diagdown .$$

(7)   (8)   (9)

Full electronic formulae such as (4)–(6) or (7)–(9) are used when we wish an explicit account of all the valency electrons—14, 14 and 12 respectively in these molecules. But these electrons are implied by the simpler formulae (1)–(3). The reader will easily acquire the facility of mentally supplying the missing electrons in such cases. Which type of formulation we use must depend on the circumstances.

We may now recapitulate by writing an electronic formula for the water molecule. First, we reckon how many outer-shell electrons are at our disposal. The number is eight: one from each H-atom and six from O. We have then to arrange these electrons so that each atom attains a stable grouping, which is an octet for the O-atom and a duplet for each H. The only reasonable solution is (10), or—with the unshared pairs 'taken as read'—(11). The ammonia molecule is similarly represented by (12) or (13). The H_2O molecule possesses two lone pairs; the NH_3 molecule one.

$$H\!-\!\overset{\displaystyle H}{\overset{|}{\underline{O}}}| \qquad H\!-\!\overset{\displaystyle H}{\overset{|}{O}} \qquad H\!-\!\overset{\displaystyle H}{\underset{}{\overset{|}{\underline{N}}}}\!-\!H \qquad H\!-\!\overset{\displaystyle H}{\overset{|}{N}}\!-\!H$$

(10)   (11)   (12)   (13)

These simple rules enable us to write appropriate formulae for a great number of covalent molecules. They apply with special simplicity so long as we confine ourselves to elements up to, and including, the Second Short Period. They cover almost all organic compounds. The octet is never exceeded in the First Short Period. It may be exceeded with elements of atomic number greater than 12, as we shall see later, but the octet remains of great importance in the Second Short Period and in the later *B*-sub-groups.

When we have written one of these electronic formulae, we have perhaps not 'explained' very much about the molecule. But at least we have placed it in a logical scheme that covers a majority of the millions of known chemical compounds. Though rough approximations, formulae such as (4)–(13) are a useful way of systematizing our knowledge as well as being a starting point for a more fundamental study of molecules.

3.5   Dative covalency

In all the electronic formulae we have considered so far, each of the atoms united by a covalent bond has provided one of the electrons for the bonding pair. We have not stressed this fact, since all the electrons in the molecule are equivalent once the molecule is formed. That it is true from an electron-accounting point of view is evident from a consideration of the formal statement,

$$|\overline{\text{Cl}}\cdot \; + \; \cdot\overline{\text{Cl}}| \; \longrightarrow \; |\overline{\text{Cl}}{-}\overline{\text{Cl}}|.$$

However, there are cases where this is not so—where one of the atoms provides both electrons of the bond. We have the transaction $A + :B \longrightarrow A{-}B$, instead of $C\cdot + \cdot D \longrightarrow C{-}D$. The $A{-}B$ bond, once formed, is covalent, like that in $C{-}D$: but, when we wish to emphasize its different ancestry, we may term it a *dative covalency**. With the same emphasis in mind, we sometimes replace the bond-line with an arrow pointing from the atom which has donated the electron-pair to that which has accepted it, i.e. $A \leftarrow B$.

When a dative bond is formed between atoms that were initially uncharged, a net separation of electrical charges results. This justifies an alternative method for drawing attention to the dative character of the bond: viz. $\overset{\ominus}{A}{-}\overset{\oplus}{B}$, in which we have added *formal charges* to the atom-symbols: positive to the donor and negative to the acceptor atom.

We may explain this more fully by the example of the union of the molecule of trimethylamine with oxygen to yield the amine oxide:

$$\underset{\displaystyle \text{CH}_3}{\overset{\displaystyle \text{CH}_3}{\text{CH}_3{-}\text{N}|}} \; + \; \overline{\text{O}}| \; \longrightarrow \; \underset{\displaystyle \text{CH}_3}{\overset{\displaystyle \text{CH}_3}{\text{CH}_3{-}\text{N}{-}\overline{\text{O}}|}}.$$

* The terms, 'coordinate link', 'co-ionic bond', and 'semi-polar bond' have also been used.

The neutral N-atom has five electrons in its outer shell, and the O-atom has six. In the amine, shown on the left side of the statement, the N-atom has a complete octet; but, for our present purpose of charge allocation, we suppose that only one electron out of each shared pair serves to neutralize the $5+$ charge on the N-core (i.e. $+7$ on the nucleus and -2 for the $1s$-electrons). On this basis, the N-atom has 2 (lone pair) $+\frac{1}{2}(6)$ (shared pairs) $= 5$ electrons. It is therefore neutral. In the oxide, on the other hand, $\frac{1}{2}(8) = 4$ electrons are counted for the N-atom, which has a net formal charge of $+1$ therefore. By a similar reckoning the O-atom, with a core-charge of $+6$, and 3 lone pairs $+1$ shared pair $= 7$ electrons, has a net formal charge of -1. That this is a rough-and-ready method of assessing the charge on an atom is recognized by the word 'formal'. But there is some degree of charge separation in such molecules. (In this case it accords with the high dipole moment of $5·0$ D; see §4.4.)

We should emphasize that the bonds we are discussing are all covalencies, and it is always permissible to represent them simply as such. We can draw attention to the dative character of the N—O bond in the amine oxide by using the formulations $(CH_3)_3N{\rightarrow}O$ or $(CH_3)_3\overset{\oplus}{N}{-}\overset{\ominus}{O}$; but the undecorated formula $(CH_3)_3N{-}O$ carries this information by implication.

Indeed a strict determination to be definitive about dative bonds can lead us into difficulties where one of the atoms carried a charge before donation of the lone pair. The ammonium ion is an instructive example. The statement,

$$3H· + ·\overline{N}| \longrightarrow H{-}\underset{|}{\overset{|}{N}}{-}H,$$

represents the formation of the ammonia molecule. From it we see that the three N—H bonds are normal covalencies. Addition of a proton to the molecule leads to the ammonium ion:

$$H{-}\underset{H}{\overset{H}{\underset{|}{\overset{|}{N}}}}| + H^+ \longrightarrow H{-}\underset{H}{\overset{H}{\underset{|}{\overset{|}{N}}}}{\rightarrow}H^{\oplus};$$

and—thinking only of this last step—we would conclude that the last N—H bond is dative, as we have in fact shown above. Once formed, however, all four N—H bonds are equivalent. (The ammonium ion has a regularly tetrahedral shape.) To mark one bond as

dative and the other three as normal covalencies is illogical. To be
sure, we might suggest that each bond is 25 % dative and 75 % normal;
but this would be too nice a description at this stage. We shall do
better to adopt the simpler expedients of either enclosing the whole
ion in square brackets and adding the charge sign, or attaching the
formal charge to the N-atom:

$$
\left[\begin{array}{c} H \\ | \\ H{-}N{-}H \\ | \\ H \end{array} \right]^{+} \quad \text{or} \quad H{-}\overset{H}{\underset{H}{\overset{|}{N}}}\oplus{-}H.
$$

A similar problem arises with the sulphate ion. The simplest
electronic formula for the unionized H_2SO_4-molecule is (14), in
which the two S—OH bonds are normal covalencies and the S → O
bonds dative, as shown. When ionization occurs by removal of

$$
H{-}O{-}\overset{O}{\underset{O}{\overset{\uparrow}{S}}}{-}O{-}H \quad (14) \qquad \left[O{-}\overset{O}{\underset{O}{\overset{|}{S}}}{-}O \right]^{2-} \quad (15) \qquad {}^{\ominus}O{-}\overset{O^{\ominus}}{\underset{O_{\ominus}}{\overset{|}{S}}}{\oplus}{-}O^{\ominus} \quad (16)
$$

the two protons, we arrive at (15), in which all four bonds are equivalent; to regard two of them as still dative would be arbitrary. Either
of the representations (15) or (16) is acceptable at this level. We shall
see later that the above discussion of sulphuric acid and the sulphate
ion is probably over-simplified (§ 3.7).

The dative bond plays an important part in the formation of
coordination compounds. To take an early example, the compound of
composition $CoCl_3 . 6NH_3$, was represented by Werner as
$[Co(NH_3)_6]^{3+} 3 Cl^-$. The chloride ions separate from the cation in
solution, but the ammonia molecules remain attached to the Co-
atom. In terms of the electronic theory, we interpret this by supposing
that the lone pairs on the NH_3-molecules each form a dative bond
with the Co-ion, which thus acquires an outermost shell of 12 electrons
when it acts as acceptor. This is a naive way of regarding coordination.
Ligand-field theory treats it with much more finesse.

3.6 Acids and bases

Another aspect of dative covalency should be mentioned. The
commonest generalization about acids and bases, and one particularly
useful to electrochemists, is that an acid is a substance whose molecule

can donate a proton, whilst a base is a substance whose molecule can accept a proton. Ordinary laboratory acids (such as sulphuric acid solution) are acids because they contain the ion H_3O^+—the so-called hydrogen ion—which can offer a proton by the procedure, $H_3O^+ \longrightarrow H_2O + H^+$. Other acids are such entities as NH_4^+, the HCl molecule, the bicarbonate ion HCO_3^-. Ordinary bases act because they contain the OH^- ion, which can accept a proton by the procedure, $OH^- + H \longrightarrow H_2O$. Other bases are the ammonia molecule and the acetate ion, $CH_3CO_2^-$. This is the *protonic concept* of acid-base behaviour.

An alternative, *electronic concept* was suggested many years ago by G. N. Lewis. It is of much wider applicability, but less useful from a quantitative point of view than the protonic concept. An acid is now a substance whose molecule can accept a pair of electrons to form a dative covalency, and a base one whose molecule can donate a lone pair. A little thought should convince the reader that this includes all the acids and bases recognized in the protonic concept. It also includes as acids such molecules as BF_3 (see §3.7) or $AlCl_3$, and as bases such compounds as aromatic hydrocarbons when they form a 'charge-transfer' complex with molecular iodine.

3.7   Electronic groupings other than the octet

A stable group of eight electrons is usually, or always, sought by the covalency-forming elements, C, N, O, F, (Si), P, S and Cl, which lie in the upper right-hand corner of the Periodic Table, as it is commonly drawn. As familiar covalent compounds—and nearly all organic compounds—involve these elements, with hydrogen, the octet is of dominant importance in an elementary treatment of valency, as we have already stressed. But other stable groupings occur.

Hydrogen we have already discussed. It has a duplet of electrons in nearly all stable compounds. Sometimes an element which can form an octet may also form moderately stable molecules in which it has an *incomplete octet*. Boron trifluoride is formed by the process,

$$|B\cdot \; + \; 3\cdot\overline{F}| \; \longrightarrow \quad \langle\overline{F}\rangle \diagdown \quad \diagup \langle\overline{F}\rangle$$
$$B$$
$$|\underline{F}|$$

Though the F-atoms complete their octets, the B-atom has only a *sextet*. Not surprisingly boron trifluoride, whilst reasonably stable, is a strong 'Lewis-acid' (see § **3.6**), and it combines readily with a base such as ammonia:

$$\begin{array}{ccc}
\text{F} & \text{H} \\
| & | \\
\text{F--B} + |\text{N--H} \longrightarrow & \text{F--B} \leftarrow \text{N--H.} \\
| & | \\
\text{F} & \text{H}
\end{array}$$

The molecule produced has a high dipole moment ($\mu = 5\cdot0$ D), as we should expect from the formal charges.

Aluminium chloride vapour at high temperatures ($\sim 800°$) also exists as $AlCl_3$ molecules with a sextet of electrons round the Al-atom. At lower temperatures the $AlCl_3$ molecules dimerize to form Al_2Cl_6 molecules:

$$2 \text{ Cl--Al} \longrightarrow \quad \text{Al} \quad \text{Al} \qquad (17)$$

The 'bridged' structure suggested in (17) is now well authenticated by electron-diffraction studies of the vapour (see § **4.5**).

Stable molecules with an atom having an incomplete octet are uncommon. There are many stable compounds which include valency groups exceeding the eight electrons of the octet. As we explained earlier, such groupings will not occur before the Second Short Period, when d-orbitals become available. Stable compounds with groups of ten and twelve electrons then occur fairly often. *A fortiori* they are likely to occur in elements of the Long Periods. The participation of d-orbitals in such circumstances is now the subject for much research, both experimental and theoretical.

We give three examples. Phosphorus pentachloride consists of PCl_5 molecules in the vapour phase. In these molecules, whose structure is explained in Table 4.3, the P-atom has a group of ten electrons in the M-shell. Accommodation can be found for these electrons by using one of the $3d$-orbitals as well as the $3s$ and the three $3p$. Sulphur hexafluoride, SF_6, is very stable and in its molecule the S-atom must have a shell of twelve electrons, which are accommodated in orbitals derived from two of the $3d$-orbitals as well as the four $3s$ and $3p$ (see § **5.5**).

In § **3.5** we represented sulphuric acid and the sulphate ion by formulae implying only an octet round the S-atom. Almost certainly some d-orbitals are drawn in here; and we may draw such formulae

$$
\begin{array}{c}
O \\
\parallel \\
O-S=O \\
\mid \\
O
\end{array} \quad (18)
$$

as (18). Any double-bond character (see §5.9) is not confined to one S—O bond; all four bonds have the same length of about 1·53 Å. A similar sharing of double-bond character may be supposed to occur in such other ions as PO_4^{3-} and ClO_4^-.

3.8   Some odd and anomalous molecules

Nearly all stable molecules, and most stable ions, possess an even number of electrons. This is good evidence for the importance of electron pairing, independent of any notions of electron spin. Molecules which break this rule are often termed *odd molecules*. A majority of recognized odd molecules are transient entities (free radicals) which have been detected, or postulated, as intermediaries in chemical reactions. For example, the methyl radical $CH_3\cdot$, where the dot signifies the unpaired electron, plays an important rôle in the reactions of many compounds. Again, the apparently simple reaction $2H_2 + O_2 \longrightarrow 2H_2O(g)$, does not take place termolecularly, but by a complex series of bimolecular steps in which such species as $H\cdot$, $OH\cdot$ and $HO_2\cdot$ are involved. The preparation of polymers, such as polythene or Perspex, requires the catalytic assistance of free radicals of this sort. The odd electron causes instability, so that a free radical generally reacts rapidly with otherwise stable molecules.

Of special interest to us here are those molecules which contrive to be moderately stable despite their odd numbers of electrons. There are not many simple examples. We must make special mention of four: nitric oxide, NO; nitrogen dioxide NO_2; chlorine dioxide, ClO_2; and some derivatives of triphenyl-methyl, $R_3C\cdot$ (where R is a phenyl group, C_6H_5, or a substituted group such as $C_6H_2(NO_2)_3$). All of these compounds are coloured—except NO—and all are reactive or very reactive. They are also paramagnetic. (The majority of compounds, with molecules whose electrons are all paired, are diamagnetic; *paramagnetism* is a characteristic of materials whose molecules have one or more unpaired electrons.)

$$
\dot{N}=O \quad (19)
$$

The NO molecule has 11 valency electrons; and it is obviously vain to attempt writing a classical, electronic bond-diagram for it. If nevertheless we persist, we arrive at some such impasse as is represented by (19) in which the N-atom has only seven electrons instead of eight.

That such molecules are rare makes them an interesting exception to our rules, rather than a serious threat. How we may try to explain these anomalies we shall discuss in Chapter 5.

Anomalous in a different way are the compounds with *electron-deficient molecules*. The simplest and best-known example is B_2H_6, which we shall use to illustrate the difficulties. As we have seen, compounds such as BF_3 or $B(CH_3)_3$ exist, with the B-atom having an incomplete octet. The corresponding hydride of boron would be BH_3. A compound of this composition is known, but its molecular formula proved to be B_2H_6, and the name 'diborane' is now given to it. At first sight, this formula suggests a resemblance to ethane, C_2H_6, and for this reason the name 'boroethane' was once used. A moment's consideration shows that this molecule lacks two electrons for the formulation (20) as an ethane analogue.

$$
\begin{matrix}
\text{H} & \text{H} \\
| & | \\
\text{H}-\text{B}-\text{B}-\text{H} \\
| & | \\
\text{H} & \text{H}
\end{matrix}
\quad (20)
$$

(21)

The problem became better defined when experimental work proved that the six H-atoms were not in fact arranged as formula (20) would require, but rather as in (21). The four H-atoms at the ends of the molecule and the two B-atoms are coplanar, whilst the other two H-atoms lie in a plane at right angles to the first plane. (Formula (21) tries to suggest this by the use of tapered bonds, more or less in the line of sight.) We shall return to these compounds later in §7.6.

Formula (21) is an example of a 'bridge' structure, and it resembles that given above for Al_2Cl_6 (which however is not electron-deficient). But there is also a compound $Al_2(CH_3)_6$, whose molecular structure corresponds to that of Al_2Cl_6, and which appears to involve 'bivalent' methyl groups in the 'bridging' positions. Some form of electron-deficiency is to be suspected.

Chapter 4

Molecular Properties and Their Measurement

4.1 Introduction

Early ideas of the chemical molecule were qualitative. The three atoms of the water molecule were believed to be linked together in the sequence H—O—H; but quantitative, or metrical, knowledge was lacking. The actual weight and size of the molecule were unknown—or known only very roughly; whether the three atoms lay in a straight line or not was unknown, as were also the interatomic distances; and there was no known measure of the strength of the bonds between the atoms.

Since 1930 physical methods for deriving information of this kind have been developed and applied with increasing accuracy and assiduity. We shall summarize these methods in § 4.5. They have had a profound influence on valency theory.

4.2 The potential-energy diagram for a diatomic molecule

We can illustrate the sort of information we are seeking, and may hope to find, by considering the *potential-energy diagram* for a molecule consisting of two atoms. This is sketched in Fig. 4.1(a), where we plot the energy of a system consisting of a pair of atoms against the distance apart (d) of their nuclei. If the atoms are involved in the formation of a stable molecule, the plot must have the general shape shown, as we can understand from the following considerations. Since there will be an attractive (covalent) force between the atoms, work must be performed when they are drawn apart; so the potential energy rises to the right. (Conversely, as the atoms approach one another from a distance, the potential energy diminishes.) However, the force of attraction is not unopposed; for, if it were, the two atoms would tend to coalesce, the energy falling to a very low value (nominally $-\infty$) as d becomes zero. On the contrary, strongly repulsive forces come into play when the atoms come close together: the positive charges on the two atomic nuclei repel one another, as do the negative charges carried by the electrons belonging to the two atoms. The repulsion, rising dramatically as d decreases, causes the curve to rise very steeply at the left-hand side. Combining both types of force, attractive and repulsive, we arrive at an overall curve of the shape shown in the Fig. 4.1(a). At a certain distance, d_e, the forces are in equilibrium, and this distance corresponds to the minimum of energy. Work must be done either to extend the bond or to compress it. Near the minimum, and for a little distance to either side of d_e, the curve is symmetrical, and in fact parabolic in shape; but it deviates more and more from symmetry as the energy rises: as the molecule is compressed the energy rises ever more steeply; as it is stretched, the energy tends to level off.

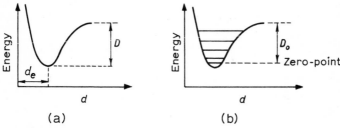

Fig. 4.1 The potential-energy diagram for a diatomic molecule.

To a first approximation we can identify two features of such a diagram with important measurable properties of the molecule. The distance, d_e, represents the normal interatomic distance. The height D represents the *dissociation energy*—the energy that must be supplied to cause the atoms to break apart. The former measures the *bond-length*, whilst the latter is one measure of the *bond-strength*.

4.3 Molecular vibrations

The view just given is oversimplified because it neglects the circumstance that the molecule is vibrating. The system of two atoms linked by a covalent bond is closely analogous to the dynamical system of two masses connected by a coil-spring. When the masses are pulled further apart and then released, the stretched spring will pull them back, and a state of vibration will become established. The frequency (ω) of this vibration depends upon the masses and upon the strength of the spring. If the vibration is of small amplitude, it obeys Hooke's law; the motion is 'simple-harmonic', and the frequency is given by the equation,

$$\omega = (1/2\pi)\sqrt{k(m_1 + m_2)/m_1 m_2},$$

where m_1 and m_2 are the respective masses and k is the *force constant*—the restoring force per unit displacement from the equilibrium position. The same formula applies to the vibration of a diatomic molecule. (The constant k for a particular molecule is an alternative measure of the strength of the bond.)

However, there are important differences between the diatomic molecule and the classical system of two masses connected by a spring. Quantum theory shows that the energy, E_v, possessed by a diatomic molecule by virtue of its vibration is given by the equation,

$$E_v = \mathbf{h}\omega(v + \tfrac{1}{2}),$$

where $\mathbf{h}$ is Planck's constant (see § 1.9) and v the vibrational quantum number, which can have integral values, 0, 1, 2, 3 ... etc. This formula emphasises two points which arise from the nature of quantum theory: first, that the vibrational energy is quantized—it can be at only the discrete levels $\tfrac{1}{2}\mathbf{h}\omega$, $1\tfrac{1}{2}\mathbf{h}\omega$, $2\tfrac{1}{2}\mathbf{h}\omega$, etc; and secondly, that there can be no level of zero vibrational energy. Even when as much as possible of the vibrational energy has been drained away, some remains; as the absolute zero of temperature is approached, some vibration persists. This residue is known as *zero-point energy*. (Its fundamental theoretical basis lies in the Uncertainty Principle (see § 2.3): if vibration were to die away completely, we would then be able to define both the position and the momentum of a particle exactly.)

The consequences of the equation can be represented by the additions we have made to the potential-energy diagram in Fig. 4.1(b). The permissible energy levels are there shown by the horizontal lines drawn across the curve. As the molecule vibrates, the point on the curve indicating its state rolls backwards and forwards across the minimum and up each side to the level of the line. The potential energy therefore fluctuates periodically, but so does the kinetic energy of the moving atomic masses, so that the total energy remains constant at the appropriate value of E_v. The state of lowest possible energy is shown by the lowest line, and not by the actual minimum of the curve. It follows that the energy of dissociation from this lowest level (D_o) is slightly smaller than the D indicated in Fig. 4.1(a). In the case of the H_2 molecule, D (calculated for $0°K$) is about 109 kcal/mole, whilst D_o is about 103, the difference of some 6 kcal being the zero-point energy.

Since a molecule is always vibrating, the exact definition of its interatomic distances or bond-lengths needs some care. One obvious way is to define the distance by d_e, the internuclear distance corresponding to the minimum of the energy curve. (The subscript e refers to the static *equilibrium* between attractive and repulsive forces.)

This would be the true distance in a hypothetical molecule without vibrational energy. The methods that are used experimentally find a distance averaged over the amplitude of the vibration, and we may call this d_o. If the potential-energy curve were strictly symmetrical and the vibration simple-harmonic in character, d_e and d_o would be the same. Because the vibration is not exactly harmonic, they differ slightly. The difference is usually small, and it need not concern us in this book; but it is of significance when distances can be measured very accurately—as they can by certain methods.

The distances between bonded atoms in a molecule are always of the order of 10^{-8} cm. This makes it convenient to express them in terms of the *Ångstrom unit*, or the *Ångstrom*, which is now defined by the relationship, $1 \text{ Å} = 10^{-8}$ cm. This unit is named after the Swedish spectroscopist Ångstrom, and the pronunciation is (approximately) 'Ongstrom'.

4.4 The dipole moment

A molecule is built up from positively and negatively charged particles—atomic nuclei and electrons; and, in a neutral molecule, the number of units of positive and negative charge are necessarily exactly equal. But what we might (rather loosely) call their respective 'centres of gravity' may, or may not, coincide. When the molecule is centro-symmetric, as it is for a diatomic molecule consisting of two atoms of the same kind, such as H_2, or O_2, or Cl_2, the centres of gravity of the positive and negative charges will coincide. When the atoms differ, they will not. In the hydrogen chloride molecule, for instance, the chlorine nucleus may be supposed to exert a more powerful attraction on the bonding electron-pair than does the hydrogen nucleus with its smaller positive charge. (Chlorine is said to possess a higher *electronegativity* than does hydrogen—see § 4.14.) The chlorine-end of this molecule therefore carries a certain residual negative charge, and the hydrogen-end an equal residue of positive charge. The amount of such separated charge is less than the unit electronic charge, such as is carried by each ion of an electrovalent compound like $Na^+ Cl^-$. The hydrogen chloride molecule, as it exists in the gaseous state, is not an ionic compound as the formula $H^+ Cl^-$ would imply. Rather it is covalent; but the bonding electron-pair is, as it were, not equitably shared. They are attracted somewhat towards the chlorine atom. We may draw attention to this by using the Greek letter delta (δ) to suggest a fractional charge and adding $\delta+$ and $\delta-$ to the usual formula as follows:

$$\overset{\delta+}{H}-\overset{\delta-}{Cl}.$$

A molecule with this property of electrical imbalance is said to be *polar*. It will tend to orientate itself in an electrostatic field, just as does a bar magnet in a magnetic field. The extent of the polarity is measured by the *dipole moment* (more fully the electric dipole moment). This moment is defined by the equation,

$$\mu = e \times d,$$

where e here stands for the effective charge—positive or negative—at each end of the molecule, and d is the distance between the centres of these respective charges. Since e is of the order of 10^{-10} e.s. unit of charge and d of the order of 10^{-8} cm, dipole moments are always expressed in *Debye units*, where $1D = 10^{-18}$ cm e.s. unit.

The actual dipole moment of a molecule can be measured experimentally, the substance being preferably in the gaseous state. The hydrogen chloride molecule, for example, has $\mu = 1.1D$, whilst a non-polar molecule like Cl_2 will have $\mu = 0$.

Dipole moments can be recognized, and measured, for molecules with more than two atoms. For example, oxygen is more electronegative than hydrogen; therefore each H—O bond in the water molecule is polar in the sense $\overset{\delta+}{H}$—$\overset{\delta-}{O}$. Were the molecule linear, the two *bond moments* (we may represent them by m) would oppose one another, giving a zero over-all molecular dipole moment. Since this molecule is in fact non-linear, they produce a resultant moment that is non-zero. The situation is suggested in Fig. 4.2(a). The two bond moments can be added vectorially, as is suggested by the 'parallelogram of forces' shown in Fig. 4.2(b). That the water molecule has an experimental dipole moment of about $1.8D$ is evidence for its non-linearity.

(a)   (b)

Fig. 4.2   Polarity of the H_2O-molecule: resolution of the overall dipole moment into two bond moments.

The above discussion will have given the impression that a non-zero dipole moment is caused simply by inequitable sharing of the bonding electron-pair. Though this is a useful way of considering bond moments and polarity in molecules generally, it has only a superficial validity (see §4.14).

4.5 Physical methods for studying molecular structure

It is beyond the scope of this book to describe the physical methods now available for determining molecular properties*. We shall attempt no more than an enumeration of the more important methods and a brief summary of their applicability, accuracy and limitations.

(a) *Spectroscopic methods.* The spectra of molecules are complicated. Though molecular spectra—especially those in the infra-red region—are constantly used in chemistry as sources of inference regarding molecular structure, a full interpretation of the spectrum is possible only for very simple molecules. When such an interpretation can be achieved, very precise information about the molecule can be derived. This information includes interatomic distances—and hence bond-lengths and -angles, force-constants for the various types of molecular vibration, the dipole moment, and sometimes the energy required to dissociate the molecule into its atoms. Accuracy is high. In favourable cases, bond-lengths can be measured to $\pm 0\cdot0002$ Å or better—a precision that calls for care in the definition of bond-length, as was explained above.

(b) *Electron diffraction by gases.* An electron of mass m moving with a velocity v is associated with a wavelength given by the de Broglie equation, $\lambda = \mathbf{h}/(mv)$ (see §2.3). Consequently a beam of electrons produces diffraction effects, just as does a beam of light or any other wave-motion when it interacts with a material, the separation of whose parts is of the same order as the wavelength. A fine beam of electrons, accelerated to a uniform velocity by a potential of about 40,000 V, has a wavelength of about 0·1 Å, which is comparable with interatomic distances in molecules. When therefore such a beam impinges on a jet of gas, and then goes on to strike a photographic plate, a diffraction pattern is recorded by the plate. This pattern consists of a series of concentric rings, the details of which are characteristic of the particular gas being studied. From a careful study of the pattern, deductions can be made about the molecular structure of the gas. This procedure is feasible only for fairly simple molecules; but, when it can be applied, interatomic distances can be determined with considerable accuracy—$\pm 0\cdot001$ Å is attainable in favourable cases. Besides having molecules that are not too compli-

---

* They are dealt with in more detail in P. J. Wheatley's *Determination of Molecular Structure* (Oxford University Press, London, 1959), and in J. C. D. Brand and J. C. Speakman's *Molecular Structure, the Physical Approach* (Edward Arnold, London, 1960).

cated, the substance must either be gaseous or be sufficiently volatile to be obtained in the vapour state.

(c) *X-Ray diffraction by crystals*. The atoms or molecules in a crystalline substance are arranged in a pattern which repeats itself regularly in three dimensions, and which therefore acts as a very effective three-dimensional diffraction grating. X-rays with a wavelength around 1 Å are convenient for crystal diffraction. The pattern, which can be recorded photographically or in other ways, is much more detailed than that given, with electrons, by a gas (because the arrangement of the diffracting material is so much more orderly). From it we can easily determine the dimensions of the repeat-unit in the crystal; and, in favourable cases, we can also determine the relative positions of the various atoms within the repeat-unit. In other words, when the crystal consists of molecules, we can determine their complete structure.

This method is considerably less accurate than the first two. Only with exceptional effort, and in specially suitable materials, can bond-lengths be found with a precision of ± 0.005 Å or better; and hydrogen atoms are harder to locate than this. On the other hand, very large molecules can be elucidated. By 1960 the structure of the molecule of vitamin B_{12} ($C_{63}H_{80}O_{14}N_{14}PCo$) had been successfully solved. Even the simpler crystalline-proteins are now beginning to yield; much of the structure of myoglobin—with about 1,200 atoms in the molecule, not counting H-atoms—is now known, though only after a prodigious effort extending over many years.

(d) *Neutron diffraction by crystals*. A beam of slow ('thermal') neutrons has an effective wavelength of about 1 Å, and will accordingly be diffracted by a crystal. The method has great technical difficulties connected with the fact that a nuclear reactor is needed to produce a beam of neutrons that is of sufficient intensity, after selection only of those within a suitably narrow range of wavelengths. The chief advantage of neutrons, from our point of view, is that they enable hydrogen atoms to be located with much greater ease and accuracy. In structural work neutron diffraction is used mainly as an adjunct to X-ray diffraction: after the structure has been determined by X-ray crystallographic methods, neutrons may be used to fix the positions of the hydrogen atoms more exactly.

There are several other physical methods for exploring the properties of molecules. Some of them are very valuable in particular circumstances, notably the various kinds of magnetic measurement. The four listed above are the methods of most general applicability.

4.6 Molecular data and the assessment of its precision

The application of these methods over the years has led to the accumulation of a great mass of information on the properties of molecules. The individual measurements are described in many thousands of original papers scattered over the literature. Most of the results have been collected, more accessibly, in various critical compilations. For our purposes, the most notable are the *Special Publications* Nos. 11 and 18 of the Chemical Society (London). We shall shortly pick out some of this information for representative molecules.

In using such compilations, the reader ought to have regard to the accuracy of a particular datum. A bond-length given as 1·473 Å will seem less impressive if its uncertainty is conceded to be ±0·015 Å. (It will be even less impressive if we find that the original researcher has—by a very human foible—been over-optimistic in assessing his own accuracy.)

Nowadays—though not necessarily in earlier work—a qualification such as '±0·015 Å' has an agreed meaning. It does not imply that true result is certain to lie within the limits stated. It is, rather, the estimated *standard deviation* of the result. This is a mildly technical matter which we need not pursue here except to state that there is a 68% chance that the true result lies between 1·458 and 1·488 Å, a 99% chance that it lies within twice these limits, and a 99·8% chance (i.e. almost a certainty) that it lies within three times these limits. All this depends on the assumption that the errors of observation are random ones and have been correctly estimated. In practice this assumption may not be completely justified.

Thus if two bond-lengths in a molecule are found to be 1·473 and 1·456 Å, each ±0·015 Å, the difference should not be regarded as significant. These results afford no sound evidence for supposing the difference to be real. But, if the standard deviations were ±0·005 Å, the difference should probably be regarded as genuine.

For simplicity, we shall not state standard deviations for the molecular dimensions that are to follow. A rough guide to the accuracy may however be taken from the number of places of decimal used. The standard deviation may be assumed to be (not far from) two or three units in the final figure given.

4.7 Some diatomic molecules

Table 4.1 lists the interatomic distances, the dipole moments, the dissociation energies and the vibrational force-constants of some

Table 4.1   Structural Properties of Some Diatomic Molecules

Molecule	d (Å)	μ (D)	D (kcal/mole)	$k(10^5 \text{dyne/cm})$
H_2	0·7413	0	104·2	5·135
F_2	1·418	0	37	~4·5
Cl_2	1·988	0	58	3·286
Br_2	2·287	0	46	2·458
I_2	2·662	0	36	1·721
O_2	1·2074	0	119	11·77
N_2	1·0976	0	226	22·96
HF	0·917	1·9	135	9·655
HCl	1·2746	1·07	103	5·157
HBr	1·412	0·8	87	4·116
HI	1·617	0·4	71	3·141
FCl	1·628	0·88	~60	4·56
CO	1·1282	0·13	256	19·02

important diatomic molecules. The reader is recommended to cultivate a 'feeling' for such data, as well as for those in later parts of this chapter. He should have a rough idea of the dimensions and other properties of the simpler molecules.

4.8   Triatomic molecules

At an elementary level, the 'structure' of a diatomic molecule has only a single feature—the interatomic distance or the bond-length; the structure is defined by a single *parameter*. The structure of a triatomic molecule (A—B—C) requires three parameters for its similar definition. These might be the three distances A—B, B—C, and $A \ldots C$. Alternatively they might be the two distances A—B and B—C and the angle A—B—C. Since in nearly all cases there are chemical grounds for supposing that A and C are each linked to B, but not to each other, we prefer the latter choice of parameters, though the former is equally valid mathematically. The structure of a triatomic molecule is therefore always described by two bond-lengths and a bond-angle (or valency-angle).

In Table 4.2 we collect such data—as well as dipole moments—for a number of triatomic molecules. When the angle is 180° the molecule is linear; and, when the atoms A and C are identical in type—as in H_2O—the two bond-lengths are equal and only one is actually given. Two triatomic ions are also included—the nitronium ion, NO_2^+, and the nitrite ion, NO_2^-—because they make an interesting series with the odd molecule NO_2.

Table 4.2 Structural Properties of Some Triatomic Molecules

Molecule	Bond-length(s) (Å)	Bond-angle (°)	Dipole moment (D)
H_2O	0·9572	104·52	1·82
D_2O	0·9575	104·47	1·85
H_2S	1·328	92	0·9
CO_2	1·161	180	<0·2
NO_2^-	1·24	115	—
NO_2	1·188	134	0·3
NO_2^+	1·154	180	—
SO_2	1·432	119·5	1·61
ClO_2	1·49	119	
O_3	1·278	117	0·5
N_2O	1·126(N—N), 1·186(N—O)	180	0·2
F_2O	1·41	103	0·3
COS	1·161(C—O), 1·560(C—S)	180	0·7
HCN	1·066(H—C), 1·153(C—N)	180	2·95
XeF_2	2·0	180	—

The vibrations of a triatomic molecule are more complex than the bond-stretching vibration of a diatomic molecule. Stretching vibrations occur indeed. There are two possible stretching 'modes': for example, in H_2O the bond-lengths may vary whilst the angle remains constant, and in one mode the lengths both increase, then decrease in phase, whilst in the other mode one length increases as the other decreases. There is also the possibility of bond-bending: the distances remain the same whilst the angle varies. A triatomic molecule has three modes of vibration* in all. (We should emphasize that the separation of the vibration into these three types is, in some ways, an artificial procedure. There is always zero-point vibrational energy at least, so that all three types are proceeding simultaneously, giving a very complex vibrational movement of the three atoms about their mean positions.)

Bond-stretching frequencies are almost always higher than those for bond-bending (or 'deformations'). In the water molecule, for instance, the two stretching frequencies are $1·13 \times 10^{13}$ and $1·10 \times 10^{13}$ sec^{-1}, while the bending frequency is $0·48 \times 10^{13}$. This

* With a linear triatomic molecule, such as CO_2, the bending mode is doubly degenerate: the molecule can be regarded as able to flex itself in two equivalent, but mutually perpendicular, planes.

difference is reflected in the corresponding force-constants provided these are expressed in comparable units. In fact this is not usually done: k for a stretching is expressed in dynes per cm; for a deformation normally as dynes per radian.

4.9   Polyatomic molecules

The structures of molecules comprising more than three atoms are a little more difficult to describe verbally. When, in an n-atomic molecule, $(n - 1)$ atoms are each linked to a single central atom, we can still define the structure in terms of bond-lengths and valency-angles. When the $(n - 1)$ attached atoms are equivalent (or, at any rate, all of the same chemical type, as in the AB_n-type molecules listed in Table 4.3), it is convenient to describe the general pattern by a descriptive adjective. Such are *planar*, *pyramidal* or *tetrahedral*; and the ascriptions we need are illustrated in Fig. 4.3. Provided the figures are regular ones, a single bond-length then usually suffices to complete the description of the molecule; but sometimes a second bond-length or a bond-angle is needed as well.

Table 4.3   Structural Properties of Some AB_n-Type Molecules

Molecule	Shape	d(A—B) (Å)	Angle (B—A—B) (°)
NH_3	trigonal pyramidal	1·017	107
NH_4^+	tetrahedral	1·02	$109\frac{1}{2}$
CH_4	tetrahedral	1·093	$109\frac{1}{2}$
CCl_4	tetrahedral	1·766	$109\frac{1}{2}$
PCl_4^+	tetrahedral	1·98	$109\frac{1}{2}$
PCl_5	trigonal bipyramidal	2·19, 2·04	120, 90
PCl_6^-	octahedral	2·07	90
SF_6	octahedral	1·57	90
SiF_4	tetrahedral	1·54	$109\frac{1}{2}$
SiF_6^{2-}	octahedral	1·71	$109\frac{1}{2}$
SO_3	trigonal planar	1·43	120
ClO_3^-	trigonal pyramidal	1·57	107
CO_3^{2-}	trigonal planar	1·31	120
NO_3^-	trigonal planar	1·243	120
BF_3	trigonal planar	1·30	120
BF_4^-	tetrahedral	1·40	$109\frac{1}{2}$
SO_4^{2-}	tetrahedral	1·44	$109\frac{1}{2}$
PO_4^{3-}	tetrahedral	~1·54	$109\frac{1}{2}$
PCl_3	trigonal pyramidal	2·043	100
$CH_3·$	probably planar	~1·1	~120
P_4	tetrahedral	2·21 (P—P)	60(P—P—P)

For example, the BF_3 molecule is trigonally planar (see Fig. 4.3): all four atoms lie in the same plane, and the three F—B—F angles are identically 120°. On the other hand, the NH_3 molecule has the shape of a trigonal pyramid. The nitrogen atom lies out of the plane defined by the equilateral triangle of hydrogens. The amount of deviation of the nitrogen atom from this plane can be expressed either by the distance ($\sim$0·4 Å), or, more commonly, by stating the H—N—H angles, which are each about 107°. (If they had been 120°, the molecule would have been trigonally planar.)

Several ions are included in the Table, as well as the unstable free radical, methyl ($CH_3\cdot$).

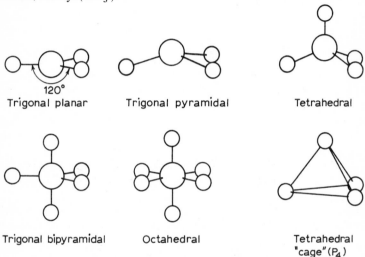

| Trigonal planar | Trigonal pyramidal | Tetrahedral |

| Trigonal bipyramidal | Octahedral | Tetrahedral "cage" (P_4) |

Fig. 4.3   Shapes of some simple molecules.

4.10   Molecular conformation

When a molecule embodies a chain of more than three atoms (e.g. A—B—C—D), a description of the structure in terms of bond-lengths and bond-angles is insufficient. We can illustrate the problem that arises by the molecule of hydrogen peroxide, H—O—O—H. The H—O distances are each about 0·97 Å, the O—O distance is about 1·48 Å, and the H—O—O angles are 105°. However by twisting one half of the molecule with respect to the other, about the O—O direction, we can vary the distance between the hydrogen atoms from about 2·73 Å when the molecule has the planar-zigzag shape illustrated in Fig. 4.4(a), down to a minimum of about 1·98 Å when it has the shape illustrated in (b). Though the bond-angles and -lengths remain constant, variation of the twist would allow the molecule to take on an infinite number of structures between these limits.

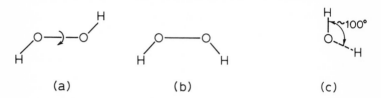

(a)   (b)   (c)

Fig. 4.4   *'Trans'* and *'cis'* conformations of the H_2O_2-molecule: (c) shows the dihedral angle actually existing in this molecule, as seen in the direction of the O—O bond.

To define the structure we need to state the *dihedral angle*. This is explained in Fig. 4.4(c). In this view we are looking at the molecule exactly along the O—O direction, so that we see these two atoms superposed. The dihedral angle is then the angle between projected directions of the two O—H bonds. In other words, it is the angle between the two O—O—H planes. In the hydrogen peroxide molecule this angle is about 100°, as shown.

This aspect of a molecule which needs to be described by a dihedral angle is known as its *conformation*. The H_2O_2 molecule requires a single such angle. A molecule with five atoms in the chain will require two dihedral angles for its full description, as well as some agreement as to how one angle is measured with respect to the other. Conformation is very important in molecules containing chains of carbon atoms, and we shall revert to this problem later (§ **4.11**).

An interesting feature of the structure suggested in Fig. 4.4(c) for H_2O_2 is that such a molecule is dissymmetric: it cannot be superposed on its own mirror-image. It is therefore formally possible to separate hydrogen peroxide into two optically active isomers, just as can actually be done with synthetic tartaric acid. However, though the molecule takes up a preferred conformation, internal rotation about a single bond occurs very easily. Interconversion of the two forms would therefore occur far too rapidly to allow of any separation of isomers.

There are three ways in which the shape of a molecule can be changed: by stretching the bonds, by bending (or deforming) them, and by twisting them: that is, by changing bond-lengths, valency-angles or dihedral angles. The useful rule is that these types of distortion occur with increasing ease. Conversely, the order of rigidities is: stretching > bending > twisting.

4.11 Some organic molecules

We now need to consider the structures of some important organic molecules. These are not conveniently presented in tabular form. Instead we shall describe them individually, with some comments.

We have seen that methane and carbon tetrachloride have regularly tetrahedral molecules, with each H—C—H or Cl—C—Cl angle about $109\frac{1}{2}°$. In the intermediate substitution products, such as methyl chloride, CH_3Cl, or chloroform, $CHCl_3$, the angles are not exactly equal (and tetrahedral), though the difference is small. Methyl chloride has (three) C—H = 1·11 Å, C—Cl = 1·781 Å, (three) H—C—H = 109·9°, and (three) H—C—Cl = 109·0°; chloroform has C—H = 1·07 Å, (three) C—Cl = 1·762 Å, Cl—C—Cl = 110·9°, and H—C—Cl = 108·9°.

Ethane, C_2H_6, has C—C = 1·534 Å and (six) C—H = 1·09 Å; the H—C—H and H—C—C angles are all very close to 109·5°. In addition this molecule poses a conformational problem. How is one methyl group orientated with respect to the other? What is the dihedral angle? At one time it was thought that there was 'free rotation' about a single C—C bond, this belief being founded on the fact that rotational isomers do not exist in such a molecule as ethane. If the rotation were literally 'free', ethane would consist of a mixture of molecules with all possible conformational shapes. In fact we now know that,

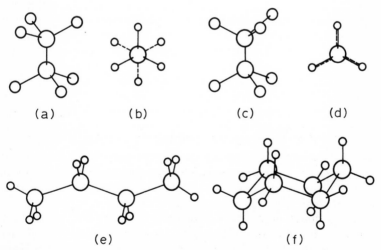

Fig. 4.5 (a)–(d) Staggered and eclipsed conformations of the ethane molecule; (e) the most stable conformation of the molecule of n-butane; (f) the cyclohexane molecule in its 'chair' conformation.

though internal rotation occurs quite easily, one particular shape is preferred. This has the *staggered* conformation suggested by Fig. 4.5(a), or in a different perspective by (b); when seen along the C—C direction, the hydrogen atoms of one methyl group lie over the gaps between pairs of hydrogen of the other group. By a twist of 120° the methyl group can pass from one preferred position to another; half-way, it passes the least favoured, *eclipsed*, conformation shown in Fig. 4.5(c) and (d). The energy difference between these two positions is about 3 kcal/mole, the staggered being the more stable by this amount.

n-Butane, C_4H_{12}, has three C—C bonds. Each of them normally has the staggered conformation, as in ethane. With butane there are three possible variants of the overall C—C—C—C chain, differing by 120° twists at the central C—C bond. The most stable is the one suggested by Fig. 4.5(e), which has a planar zigzag of carbon atoms, with the maximum possible separation of the end-atoms consistent with the maintenance of approximately tetrahedral valency-angles. This is energetically the most stable form for all long chains of singly bonded carbon atoms. The polymethylene chains in the molecules of fats and fatty acids are thus extended in the crystalline state. However, by sometimes adopting one of the other dihedral angles at one or other of its C—C bonds, such a chain can easily flex itself, and no doubt does so in solution, or in the gaseous state.

When a chain of six CH_2-groups joins up with its own end, we get the ring-molecule of cyclohexane, C_6H_{12}. The most stable conformation of this molecule is the so-called 'chair' form sketched in Fig. 4.5(f). All the C—C bonds have staggered arrangements, and the C—C distance is 1·54 Å.

Diamond is not a molecular substance in the ordinary sense of the term (see § 1.7), but its structure has relevance here. Each carbon atom is bonded covalently to four others, so that the whole crystal can be seen as a giant-molecule. The valency-angles are all exactly tetrahedral, and the C—C distance, which is very accurately known, is 1·5445 Å at 18°C. The conformation about each C—C single bond is staggered. Indeed we can, in the diamond structure, make out an infinite array of cyclohexane-type rings. The crystal can be regarded as an aliphatic molecule extending in all directions.

The simplest unsaturated hydrocarbon is ethylene, C_2H_4. The double bond does not allow one CH_2-group to twist easily with respect to the other. The two groups lie in the same plane, with C—H = 1·086 Å, C=C = 1·336 Å, H—C—H = 117°, and H—C—C = 121°.

The acetylene molecule is linear, H—C ≡ C—H, with C—H = 1·06 Å, and C≡C = 1·205 Å.

The molecule of benzene, C_6H_6, proves to have the regular, flat-hexagonal shape predicted by Kekulé's original theory. The six C—C bond-lengths are all 1·397 Å and the valency-angles all 120°; the C—H distance is 1·08 Å. In most aromatic compounds the C—C distances in the benzenoid rings are about 1·40 Å, but distinct variations occur when the carbon atoms are not all equivalent, as they are in the benzene molecule. In the molecule of naphthalene, for example, the C—C distances range from 1·36 to 1·42 Å (and in a way which—in this case—theory can partly explain, as we shall see in § **5.9**), whilst the C—C—C angles range from 119° to $121\frac{1}{2}°$.

In the molecules of carboxylic acids of the type

$$R-C\underset{OH}{\overset{O}{\diagup\diagdown}}\,,$$

where R is some organic radical, the C—O distances differ: to the hydroxylic oxygen atom it is about 1·30 Å, and to the carbonyl oxygen about 1·20 Å. On the other hand, in the anion of a carboxylic acid salt, the C—O distances are equal at about 1·25 Å. In the molecule of methyl alcohol, the C—O(H) length is decidedly longer than that in the acid: it is 1·43 Å. In a nitro-compound, $R-NO_2$, the N—O lengths are equal at 1·21 Å.

4.12   Covalent bond radii

As soon as a mass of data on molecular properties becomes available, it is natural to look for evidence of systematic regularities. Such guiding principles prove to be most evident in covalent bond-lengths. Two questions may be asked: Is the bond-length between two atoms of given types constant? Is the bond-length additive for the two types involved?

A casual look at the data in § **4.11** would suggest that the answer to the first question must be 'No'; for the distances between the carbon atoms in ethane, ethylene and acetylene are about 1·54, 1·34 and 1·21 Å respectively, differing far beyond their standard deviations. However these are bonds of different multiplicities. We ought to confine attention to bonds of the same order, e.g. single bonds. If we do this, and if we avoid certain special cases which we shall discuss in the next chapter, then the answer becomes, 'Yes, or nearly constant.' Single bonds between carbon atoms are usually in the range 1·52–1·55 Å; C—Cl mostly between 1·73 and 1·77 Å; C—H bonds around 1·1 Å. The rule of constancy is good enough to be useful.

What we mean by the word 'additive' in the second question needs some explanation. If we regard the atoms as spheres which, when the atoms form a covalent bond, just come into contact, then the distance between their nuclei would be the sum of the radii of the two spheres. So our question can be re-phrased thus: can we attribute to each sort of atom a characteristic radius, r, such that when two atoms A and B combine covalently $d = r_A + r_B$? Put thus, the question at once suggests a means for testing it. Consider the three single covalent bonds, A—A, B—B and A—B. If the equation holds, we should have:

$$d(A—B) = r_A + r_B,$$

$$d(A—A) = r_A + r_A, \quad \text{and}$$

$$d(B—B) = r_B + r_B;$$

whence it follows that

$$2d(A—B) = d(A—A) + d(B—B).$$

The length of the A—B bond should be the mean of those for A—A and B—B. Reference to the data will provide examples of this. The simplest case is where A = carbon and B = chlorine, where

$$d(C—C) = 1·54 \text{ Å}, \quad \text{and} \quad d(Cl—Cl) = 1·98 \text{ Å};$$

the mean of these is 1·76 Å, which agrees well with an average value for C—Cl bonds.

The values of r needed to suit this scheme are known as *covalent radii*. Acceptable values for the elements which commonly engage in covalent bonds are given in Table 4.4, which is based on a table due to Pauling.

Table 4.4   Covalent Bond Radii (Å)

H	B	C	N	O	F
0·30	0·88	0·77	0·70	0·66	0·64
		0·665	0·60	0·55	(double bond)
		0·60	0·55	0·50	(triple bond)
		Si	P	S	Cl
		1·17	1·10	1·04	0·99
				Se	Br
				1·17	1·14
					I
					1·33

Double, or triple, bonds are usually restricted to the atoms of the first short period, and normally to carbon, nitrogen and oxygen. The table includes suitably smaller radii for these elements, applicable when they form multiple bonds.

We should perhaps repeat that the additivity principle does not hold exactly. But within its limitations it applies approximately. Like the rule of constancy, it is therefore useful.

4.13   Bond-energy terms

We may next ask whether similar generalizations obtain for covalent bond-dissociation energies. Is the amount of energy absorbed when (say) a C—H bond is broken always the same, irrespective of the nature of the remainder of the molecule? And, if it is, can it be expressed as the sum of contributions from each atom? The short answer to the second question is 'No'. To the first, a qualified 'Yes' can be given, but this needs some explanation.

We start with methane as an example. We first wish to know the heat evolved ($-\Delta H$, at constant pressure) when a mole of the gas is formed from its elements in their normal forms:

$$2\,H_2 + C(s) \longrightarrow CH_4(g). \tag{1}$$

It is almost always experimentally easier to measure the heat of combustion:

$$CH_4(g) + 2\,O_2 \longrightarrow CO_2(g) + 2\,H_2O(l); \quad -\Delta H = 212{\cdot}9 \text{ kcal/mole.}$$

Given also the heats of combustion of graphite, $C(s)$, to CO_2 and of hydrogen to $H_2O(l)$, $-\Delta H = 94{\cdot}05$ and $68{\cdot}32$ kcal/mole respectively, we can easily deduce that the heat of formation of methane, according to equation (1), is $-\Delta H = 17{\cdot}9$ kcal/mole. This reaction involves not only the formation of four C—H bonds, but also the breaking of two H—H bonds, as well as of the bonds holding the carbon atom in graphite. So we need to use the additional experimental data represented in equations (2) and (3).

$$2\,H \longrightarrow H_2; \quad -\Delta H = 104{\cdot}2 \text{ kcal/mole,} \tag{2}$$

$$C(g) \longrightarrow C(s); \quad -\Delta H = 171{\cdot}7 \text{ kcal/mole.} \tag{3}$$

Elementary manipulation of the three thermochemical equations, (1), (2) and (3), leads to the result:

$$C(g) + 4\,H \longrightarrow CH_4(g); \quad -\Delta H = 398{\cdot}0 \text{ kcal/mole.}$$

This is the amount of heat liberated, at 25°, when the four C—H bonds of the methane molecule are formed from the atoms, in a constant-pressure reaction. As the bonds appear to be equivalent, the 398·0 can reasonably be divided by four, to yield 99·5 kcal/mole, which can be regarded as the C—H *bond energy* or its *bond-energy term*.

(It does not follow that 99·5 kcal would be the energy needed at each step if we were to dismantle the CH_4 molecule to give successively CH_3, CH_2, CH, and C. The respective dissociation energies are about 102, 90, 124 and 82 kcal; the values vary as, after each cut, the residue of the molecule reorganizes itself. But, necessarily, the sum of these values must add up to 4 × 99·5 kcal/mole.)

We proceed to ethane, whose heat of combustion is $-\Delta H = 372\cdot9$ kcal/mole, whence—as the reader should verify—we find:

$$2C(g) + 6H \longrightarrow C_2H_6(g); \qquad -\Delta H = 676\cdot2 \text{ kcal/mole.}$$

This heat evolution accompanies the formation, from atoms, of six C—H bonds and of one C—C bond. There is no unambiguous way of allocating the 676 kcal amongst the seven bonds; but, if we care to make the not unreasonable assumption that the C—H bond-energy term found in methane also applies in ethane, we are left with $676\cdot2 - 6 \times 99\cdot5 = 79\cdot2$ kcal/mole for the dissociation energy of the C—C bond, or its bond-energy term.

Proceeding in this way, we try to produce a consistent set of bond-energy terms that would apply over a whole range of different compounds. Strict self-consistency would require that a constant term would cover a particular kind of bond in any molecule. If this were true, two isomeric molecules with exactly the same numbers and types of bonds would have exactly the same value for their heats of formation, or combustion. This is not so: for example, the heat of formation of neopentane, $(CH_3)_4C$, is 4 kcal greater than that of *n*-pentane, $CH_3 . CH_2 . CH_2 . CH_2 . CH_3$. However, making some concessions, we can arrive at a set of terms which enable us to attain fair agreement between the observed and calculated energies of formation of many gaseous molecules from their atoms. The agreement is good enough in most cases to show up the importance of certain exceptions which we shall discuss in Chapter 5. A suitable set of bond-energy terms is given in Table 4.5.

Table 4.5   Bond-Energy Terms (kcal/mole)

H—H	H—C	H—N	H—O	H—F
104	99	84	110	135
C—C	C=C	C≡C	C—O	C=O
83	146	~200	86	178
C—N	C=N	F—C	Cl—C	H—Cl
73	147	116	81	103

4.14   Electronegativity and bond moments

The covalent bond-length is roughly constant for a bond between two given types of atom, and it can be roughly expressed as a sum of bond radii. Something to the same general effect can be stated about bond energies. We may speculate whether any corresponding principles apply to bond dipole moments. We do not need to present the detailed evidence here, but the conclusions are as follows.

In a rough-and-ready way only, the electric moment associated with a particular sort of bond can indeed be taken as constant. This is not exactly true, as the rest of the molecule may have a decided effect.

We would not expect the bond moment to be a *sum* of effects due to each atom. Since the moment in a simple-minded view depends on the competing attractions exerted by the two atomic cores upon the bonding electron-pair, we should, rather, expect it to be the *difference* of some property of the two atoms. This property is the *electronegativity* (see § **4.4**). Some of Pauling's set of electronegativities are displayed in Table 4.6.

Table 4.6   Electronegativity Values (Pauling)

H	2·1						
C	2·5	N	3·0	O	3·5	F	4·0
		P	2·1	S	2·5	Cl	3·0
				Se	2·4	Br	2·8
						I	2·5

Two examples will help to explain the use of such a table. The difference between the Pauling electronegativities of bromine and hydrogen is $2·8 - 2·1 = 0·7$, which happens to agree well with the experimental value of $0·8$ D for the dipole moment of HBr. For a diatomic molecule this is the same as the conventional bond moment. The difference between oxygen and hydrogen is $1·4$, which should be comparable with the bond moment (in D) of the H—O bond, the oxygen atom being negative. As we saw in Table 4.1, the water molecule has an overall dipole moment of about $1·82$ D, and the valency-angle at the oxygen atom is $105°$. If we look upon the $1·82$ as the vector sum of individual moments along each bond, we can calculate a value for the bond moment, m, from the observed dipole moment as follows (see Fig. 4.2):

$$\mu = 2m \cos (105°/2); \qquad m = 1·82/(2 \cos (52\tfrac{1}{2}°))$$

$$= 1·50 \text{ D.}$$

The similarity to the electronegativity difference is typical.

Electronegativity, regarded as a measure of the tendency of an atom to attract towards itself the electron-pair which bonds it to another atom, is a useful concept in many branches of chemistry. However, the reader should be warned not to take the interpretation of bond-polarity, suggested by the above discussion, too literally. Our discussion will have suggested that the bond moment arises simply because the pair of electrons is not shared equitably. Fairly detailed wave-mechanical calculations have now been done for simple molecules like HCl; and these make it appear that the other electrons are heavily involved too. The lone-pair electrons on the chlorine atom probably have at least as much influence in producing the polarity as do the shared pair.

4.15   Ionic radii

In a typically electrovalent compound, like crystalline sodium chloride, we cannot distinguish a discrete Na—Cl molecule. However the distance between the sodium ion and each of its six equidistant, neighbouring chloride ions can be accurately measured by X-ray diffraction. It proves to be 2·814 Å. Corresponding interionic distances are known for many salts. Analogously to covalent radii for covalent bonds, we can derive a system of *ionic radii* for ionic solids. They are sometimes known as *crystal radii*. Their use is less straightforward, but the concept of ionic radius is useful. Some values are listed in Table 4.7.

Table 4.7   Ionic (Crystal) Radii (Å)

		Li^+		
		0·60		
O^{2-}	F^-	Na^+	Mg^{2+}	Al^{3+}
1·40	1·36	0·95	0·65	0·50
S^{2-}	Cl^-	K^+	Ca^{2+}	
	1·81	1·33	0·99	
	Br^-	Rb^+	Sr^{2+}	
	1·95	1·48	1·13	
	I^-	(NH_4^+)	Ba^{2+}	
	2·16	~1·49	1·35	

In the sequence S^{2-}, Cl^-, (Ar), K^+, Ca^{2+}, the number of electrons is constant at 18. The steady diminution of ionic radius is therefore to be expected: as the nuclear charge increases, the electrons are pulled more closely towards the centre. Though not an ion, the argon atom falls logically into the series, and its atomic radius is, appropriately, about 1·54 Å.

4.16   Van der Waals radii

The two chlorine atoms in the Cl_2 molecule, joined by a strong covalent bond, are about 1.98 Å apart. But the atoms of different molecules cannot get anywhere nearly as close together as this. Two molecules begin to repel one another strongly whenever their respective atoms come within about 3.6 Å. If this were not so, the Cl_2 molecule would lose its identity. The minimum distances to which non-bonded atoms can approach each other can be found by various methods, and they can usefully be attributed to a *van der Waals radius* for each kind of atom. (The name is taken from the well-known equation of state, in which the parameter b in the $(V - b)$ factor takes account of the space actually occupied by the molecules of a gas.) Values for some of the elements are listed in Table 4.8.

Table 4.8   Some van der Waals Radii (Å)

H	1·2	N	1·5	O	1·40	F	1·35
S	1·85	Cl	1·80	Br	1·95	I	2·15
Methyl group ∼2·0				Aromatic ring ∼1·8 (normal to plane)			

The principle is that powerful repulsive forces begin to come into operation whenever two non-bonded atoms—whether they belong to the same molecule or not—approach one another to within a distance equal to the sum of their van der Waals radii. These repulsive forces are often described as *steric*. An example may help. The molecule of nitrobenzene (1) and the anion of benzoic acid (2) each have their atoms coplanar, or very nearly so; the NO_2- or CO_2^--groups are in the plane of the ring. At first sight, we might expect the same to hold for the anion of *o*-nitrobenzoic acid (3). But, if this were so, two non-bonded atoms (marked by asterisks) would come within a distance of about 1.5 Å, which is far less than the van der Waals distance of $2 \times 1.4 = 2.8$ Å. To avoid the stresses which would result, the carboxylate group is in fact turned through some $70°$ out of the plane of the other atoms.

Though this concept is useful, the van der Waals distance is not a precisely defined quantity. In special circumstances some infringement of the formal limits is tolerated. For example the Cl-atoms in a molecule of carbon tetrachloride are not directly bonded to each other; and, as the reader should be able to work out from the information given in Table 4.3, the distance between any pair is only $\sqrt{8/3} \times 1\cdot76 = 2\cdot86$ Å. This is much less than the $3\cdot6$ Å demanded by Table 4.8. The stabilization of the molecule resulting from the formation of four strong C—Cl bonds offsets any stress due to the proximity of pairs of non-bonded atoms. In other words, the van der Waals radius of an atom may be reduced in a direction lateral to that in which it has formed a covalent bond.

4.17   Molecular models, their uses and limitations

Since a molecule is three-dimensional, models are useful for enabling us to visualize its shape. With a complex molecule a model is often essential. On the other hand, solid models are bulky; and certainly, in a book, we have to get by as best we may with two-dimensional representations. An impression of the third dimension can of course be conveyed by printing a pair of stereoscopic views of the model. (It is now practicable to programme an electronic computer so that, given the atomic coordinates, it will produce a pair of scaled stereo-drawings.) But then the reader, unless he be an 'optical athlete' needs to look at the diagram through a special viewer. Alternatively, we can try to suggest three dimensions in a single drawing by such devices as 'tapered bonds' (e.g. § **3.8**, formula (21)).

Though we cannot use molecular models in a book, the chemistry student will make increasing use of models as he progresses. It is relevant to the purpose of this primer to consider the 'ethics' of molecular models.

The simplest, and cheapest, type is the 'ball-and-spoke' model. We start with cork, or plastic, or even plasticine, spheres, which may perhaps be coloured according to some code for the various elements. These spheres are bored with suitably disposed holes, and metal rods can be fitted into these holes to represent the bonds which join atoms together. Fig. 4.6(a), for instance, gives an impression of a ball-and-spoke model of the water molecule. The lengths of the spokes can be chosen so that the model is to scale—say, 1 in. = 1 Å; and the holes are bored so as to give the correct bond-angle. Many of the diagrams of molecules used in this book can be regarded as sketches of a ball-and-spoke model. In a sense, all structural formulae are highly conventionalised symbols of such models.

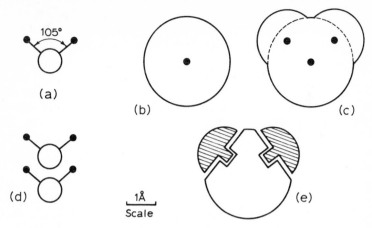

Fig. 4.6 Models illustrating some properties of the H_2O-molecule.

A ball-and-spoke model fails to convey the flexibility of a molecule or its ever-present vibrations. More fundamental limitations become evident when we try to 'consider what a molecule is really like'. (The implied question is of course unanswerable; for the senses of sight and touch, which can be used to examine models, are totally inapplicable to molecules.) Nearly all the mass of a molecule resides in the atomic nuclei, which would be invisibly small on the scale we have suggested. The balls are much too large to represent the nuclei. And they are too small to represent the whole atom. An oxygen ion, O^{2-}, has a spherically-symmetrical electron-density cloud, which dies away gradually without any definite boundary, but which becomes almost negligible at a distance of about $1\cdot4$ Å from the centre—the van der Waals radius of oxygen (Fig. 4.6(b)). When two protons are added to O^{2-}, to produce a water molecule, they imbed themselves in the electron-density sphere, each at a distance of about $1\cdot0$ Å from the centre. This causes a localized condensation of density round each; and, remembering the $1s$-orbitals of the H-atoms, we must add a van der Waals radius of about $1\cdot0$ Å about each proton. The effect we wish to suggest is shown in Fig. 4.6(c). This diagram also reminds us that it would be impossible to pack two water molecules as closely together as the ball-and-spoke model might allow us to suppose (Fig. 4.6(d)).

To match Fig. 4.6(c), a more sophisticated type of model was invented by H. A. Stuart, and has since been developed by others. The atoms are based on plastic spheres with radii proportional to the van der Waals radius of the particular element. The plastic is cut away in suitable directions, and the flat surfaces thus produced are supplied with holes or pegs, so that two bonded atoms can be fitted together with their centres separated by the appropriate covalent bond-length. The idea is suggested by Fig. 4.6(e). This type of model is more realistic when we wish to consider how molecules may be packed together. On the other hand, we no longer have an unobstructed view of the bonding skeleton of the molecule. For this reason, as well as for their simplicity, ball-and-spoke models are most generally useful. But, when we use them, we must remember their conventionality.

Chapter 5

The Wave–Mechanical Interpretation of Valency

5.1 Survey of the problem

The force of attraction between positively and negatively charged bodies is demonstrated at an early stage of our scientific education; and, therefore, for most of us, it becomes a familiar phenomenon, though it is no less 'mysterious' than other features of our universe. This familiarity makes us feel that we 'understand' electrovalency—the strong attraction between oppositely charged ions. It is due to a force well-recognized in classical physics.

Covalency is more obviously mysterious. Classical theory cannot explain why the sharing of a pair of electrons (whatever that means) may lead to an attractive force between two atoms just about as strong as electrovalency. Wave mechanics has led to an interpretation of covalency; and this achievement in 1927 had a profound influence on chemistry.

At that date Heitler and London made use of a wave-mechanical effect, not known to classical physics, to calculate the binding energy in an H_2-molecule. Their result was not in exact agreement with the experimental value, but they were able to account for some 70% of the energy. Later, when more detailed calculations were made, almost exact agreement was attained.

Even for so relatively simple a system as H_2, the mathematics is cumbersome, and progress can be made only by approximate methods, which become laborious when some precision is sought. For larger molecules the situation is much more difficult. However the successful treatment of H_2 gives us some confidence that we understand the covalent bond in principle.

Some mathematical elaboration is essential for a proper appreciation of wave mechanics. Its application to valency theory is dealt with in several excellent monographs*. However this primer tries to

* Notably: C. A. Coulson, *Valence*, 2nd edn., Oxford University Press, London, 1961; J. W. Linnett, *Wave Mechanics and Valency*, Methuen, London; Wiley, New York, 1960; C. W. N. Cumper, *Wave Mechanics for Chemists*, Heinemann Educ., London, 1966; H. B. Gray, *Electrons and Chemical Bonding*, Benjamin, New York, 1964.

avoid mathematics. Its plan is to give a descriptive account of two approximate methods that can be applied to the H_2-molecule, and thus to introduce a number of useful concepts that help to enlarge our general understanding of valency. This can be useful in a preliminary survey. The reader who wishes to study valency seriously will have to proceed in due course to a mathematically-based presentation.

In principle wave mechanics can account for all the observable properties of any collection of nuclei and electrons i.e. of any molecule. In practice mathematical difficulties preclude the rigorous solution of any system involving more than two bodies. Hence even for so simple a chemical system as H_2, which has four bodies, approximations are needed. For more complicated molecules even more drastic approximations are needed. But the situation is not as discouraging as this might suggest; for, by a rather elaborate set of approximations, we can account for the key properties of simple molecules, like H_2, almost exactly, whilst, with the more complicated molecules, we can use wave-mechanics to relate the values of some property of a series of molecules in a useful way, without being able to calculate absolute values for this property. In such ways our understanding of molecular structure has been immensely extended.

5.2 The variation method

Let Ψ stand for the unknown—and certainly very complicated— wave-function that represents everything we need to know about the electrons in some molecule. Suppose we then take two simpler wave functions ψ_1 and ψ_2, each of which we hope might approximate to Ψ. Since ψ_1 and ψ_2 are relatively simple, we can calculate the energies, E_1 and E_2, that would be possessed by the hypothetical molecules represented by ψ_1 and ψ_2 respectively. The Variation Principle tells us that, if E_1 is less than E_2, then ψ_1 is a better approximation to Ψ than is ψ_2.

We can go further. Suppose we now form a *linear combination* of the two approximate wave-functions, $c_1\psi_1 + c_2\psi_2$, where c_1 and c_2 are coefficients specifying the relative amounts of ψ_1 and ψ_2 that are to be mixed. We can then calculate the energy for the mixed function and find how this varies as we vary c_1/c_2. This energy will be lower than either E_1 or E_2; and the ratio c_1/c_2 corresponding to the minimum energy will represent the best approximation to Ψ we can achieve by mixing ψ_1 and ψ_2, and a better approximation than either separately. Mixing is not confined to two functions. We can similarly minimize the energy for $c_1\psi_1 + c_2\psi_2 + c_3\psi_3 + \dots$

When $\psi_1, \psi_2, \ldots$ are chosen judiciously, a reasonably satisfactory approximation can usually be reached, as is indicated by a close approach of the calculated energy to that observed. This Variation method is often used in the treatment of molecules.

5.3   The hydrogen molecule

A theoretical approach to the H_2-molecule faces two obvious challenges: to calculate values for the energy of dissociation, D, and the internuclear distance, d_e (see Fig. 4.1). From Table 4.1 we find that the experimental values are 104 kcal/mole (or 110 if we include zero-point energy) and 0·74 Å respectively. Two methods have been applied, each using the Variation Principle. They are the *molecular-orbital* method (which we shall describe first), and the *valency-bond* method.

An atomic orbital (a.o.) is centred on a single nucleus. Its classical counterpart is the orbit of a single electron revolving round its nucleus analogously to a planet revolving round the Sun. A molecular orbital (m.o.) describes the situation of an electron which moves under the influence of two or more atomic nuclei. It corresponds to the problem of a planet moving in the gravitational field of a double star—a dynamical problem which cannot yet be solved rigorously.

When the electron is close to the first nucleus, its m.o. will resemble an appropriate a.o. of the first atom; when it is close to the second nucleus, its m.o. will resemble an a.o. of the second atom. This suggests that the m.o. might be approximated by combining two a.o.; and in the LCAO method (LCAO = linear combination of atomic orbitals) this is done by the Variation procedure. In the case of the H_2-molecule we form an m.o. by combining two a.o. which for the H-atoms in their ground states will be $1s$-orbitals:

$$\Psi(\text{m.o.}) = c_1\psi(1s)_1 + c_2\psi(1s)_2.$$

Since the H-atoms are wholly equivalent, the numerical values of c_1 and c_2 must be the same. Remembering that it is ψ^2 which represents electron density, we see that $c_1^2 = c_2^2$, or that $c_1 = \pm c_2$. For our purpose both c s can be taken as unity, so that we have two m.o.:

$$\psi(1s)_1 + \psi(1s)_2 \quad \text{and} \quad \psi(1s)_1 - \psi(1s)_2.$$

By combining two a.o. we have produced two m.o.

To see what this means for two H-atoms close enough together to form a covalent bond we can make use of formal diagrams like Figs. 2.3(d) and 2.4(d). On this convention Fig. 5.1(a) represents two

H-atoms whose 1s a.o. overlap appreciably. As ψ (1s) implies a stationary wave whose displacement is alternately positive and negative, the m.o. with the positive sign means that the waves of each atom are in phase (Fig. 5.1(b)); the m.o. with a negative sign that they are out of phase (Fig. 5.1(c)). Since in the latter case the a.o. are exactly equal in magnitude and opposite in sign, they must cancel one another in a nodal plane midway between the atoms, and indicated by the broken line in Fig. 5.1(c). Hence electron density will be low, or zero, between the nuclei. The two atoms will then repel, rather than attract, one another. No stable molecule will be formed. On the other hand, when the waves combine in phase, the situation suggested in Fig. 5.1(b) develops; there is some concentration of electron density between the nuclei. A net force of attraction

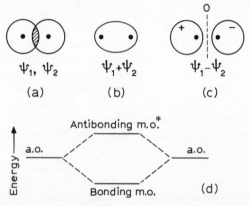

Fig. 5.1   Overlap of two 1s-orbitals to yield (b) a bonding molecular orbital, and (c) an antibonding molecular orbital.

results. So, after all, the attraction arises from the familiar electrostatic effect: the protons are held together by the negative charge between them. What is new—and was not recognized in classical physics—is that their wave-nature makes it possible for the electrons to behave in this (otherwise improbable) way.

Thus the linear combination of two a.o. leads to two m.o., one with higher energy, the other with lower energy than that of two separate H-atoms. The former is said to be *anti-bonding*, and when the two electrons are accommodated in it, there is a repulsive force between the atoms; the latter is *bonding*, and, when the electrons are in this m.o., there is an attraction. Figure 5.1(d) symbolizes the situation, the asterisk being a conventional device for marking the antibonding orbital. We note that Pauli's Principle holds: a single m.o. can accommodate only two electrons. Calculations with this m.o. treatment enable us to derive values for D and d_e of 62 kcal/mole and 0·85 Å, which are of the correct order. Various corrections ought to be applied, and they greatly improve the agreement.

We shall look at some large molecules from this m.o. viewpoint later. First we must sketch the alternative approach to the H_2-molecule—the valency-bond (v.b.) approximation. We again start with two hydrogen atoms, H(1) and H(2) each with its electron, as they come into proximity. Let ϕ_1 represent the initial situation of two close, but non-communicating, atoms.

$$\phi_1, \; H(1)e(1) \, H(2)e(2); \quad \phi_2, \; H(1)e(2) \, H(2)e(1).$$

Calculation shows that two such atoms exert only a weak attraction on one another. However it is a principle of wave mechanics that electrons are indistinguishable; so the alternative situation, ϕ_2, is equally relevant. We may expect a better approximation to the whole molecule (Φ) if we use a linear combination of ϕ_1 and ϕ_2:

$$\Phi(\text{v.b.}) = c_1\phi_1 + c_2\phi_2.$$

Here too, because of the equivalence of ϕ_1 and ϕ_2, the c s will be numerically equal, though they may have the same, or opposite, signs. Provided we use the positive combination, $\phi_1 + \phi_2$, we find an enhanced attraction between the atoms. Using this method in 1927, Heitler and London found $D = 83$ kcal/mole and $d_e = 0·87$ Å. This first estimate was improved by applying various corrections.

The formal treatment by the Variation method looks to be identical with that used in the m.o. method; and this is so. The approach is basically different, as we have tried to emphasize by using different symbols. Whereas Ψ stands for a m.o., which may accommodate two electrons, Φ stands for the wave-function for the whole molecule. The difference between m.o. and v.b. methods will become more obvious when we apply them to some larger molecules (see **§5.8**).

The main source of the covalent force, according to this Heitler–London theory, arises out of the indistinguishability of the electrons, so that both ϕ_1 and ϕ_2 are necessary. There is sometimes said to be an 'exchange force' between the atoms because the electrons have been exchanged as between ϕ_1 and ϕ_2. However we must not suppose anything so naive as that the electrons are literally exchanging places. In Coulson's words, $\Phi(\text{v.b.})$ corresponds to an attractive force because it allows 'either electron to be near either nucleus (and so) gives rise to a lowering of energy'. The electrons in the two atoms must have opposite spins, and the term *spin-coupling* is sometimes applied to this treatment.

5.4   Molecular orbitals in some larger molecules

We now apply the m.o. concept to some more complicated molecules. Detailed calculations are often difficult or impossible here; but the quasi-pictorial notion of m.o. formed by the in-phase overlap of suitable a.o. is helpful and illuminating.

We begin with the HF-molecule. The H-atom has its electron in the $1s$-orbital. Fluorine has the electronic structure $1s^2\,2s^2\,2p^5$; the unpaired electron is in one of the $2p$-orbitals of the F-atom. We can represent the approach of these two atoms by Fig. 5.2(a), where the H-atom is supposed to approach in such a direction that its s-a.o. overlaps with one lobe of the p-a.o. of the F-atom. Two m.o. can result from the overlap: the out-of-phase combination gives a node between the nuclei and is antibonding; the in-phase overlap builds up electron density between the nuclei and is bonding. This orbital holds both electrons of the bond, and it is represented in Fig. 5.2(b).

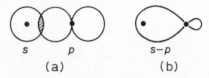

s p $s-p$

(a)   (b)

Fig. 5.2   Formation of a σ-bond by overlap of an s- and a p-orbital (HF).

This m.o. has the property of being cylindrically symmetrical about the bond-direction: if either of the atoms is rotated about the H-F axis, the amount of overlap of the a.o. is unaffected. Such an m.o. is denoted by the Greek letter sigma (σ). The bond formed when two electrons enter this m.o. is known as a σ-bond.

The formation of a water molecule is represented in a similar way in Fig. 5.3. The O-atom has two unpaired electrons, each in a $2p$-orbital. Though these a.o.s are equivalent, we represent one of them by broken lines for clarity. Each orbital may overlap with the $1s$-a.o. of an H-atom; and, provided the phases are favourable, two bonding m.o. result, as shown in Fig. 5.3(b). The four valency electrons are accommodated in these two m.o., giving rise to two σ-bonds.

There is a stereochemical implication here. Since the two p-orbitals are mutually perpendicular, the σ-bonds will tend to subtend an angle of 90°. The actual bond-angle in H_2O is distinctly greater than

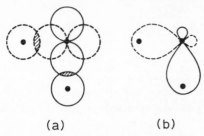

(a)   (b)

Fig. 5.3 Formation of the H_2O-molecule.

90° (see Table 4.2). The simplest way of explaining this increase is to suppose that the angle is opened up by repulsive forces between the two sets of bonding electrons.

The formation of a pyramidal NH_3-molecule from an N-atom, with three unpaired electrons, each in a singly occupied $2p$-a.o., can be similarly explained.

5.5 Hybridization of atomic orbitals

We have just given an account of the molecules HF, H_2O and H_3N in terms of m.o. resulting from the combination of a $2p$-a.o. of the larger atom with a $1s$-a.o. of an H-atom. This procedure fails if we try to apply it to methane, CH_4. The C-atom in its ground state has the structure $1s^2\ 2s^2\ 2p^2$: there are two unpaired electrons in the p-a.o., and these might be supposed to give rise to the methylene molecule CH_2. Certainly the ground-state atom is only bivalent. The atom could be made potentially quadrivalent by 'promoting' one of the $2s$-electrons to a $2p$-orbital: $1s^2\ 2s^1\ 2p^3$. This step would require an input of around 100 kcal/mole, which might 'pay off' if the atom could then form four covalent bonds. There could now be overlap between each p-a.o. and the $1s$-a.o. of an H-atom, leading to three bonding m.o., and so to three strong σ-bonds (C—H), mutually perpendicular; but the remaining $2s$-a.o. of the C-atom could overlap only very inadequately with the $1s$-a.o. of the fourth H-atom; and only a weak bond would be expected.

We know that carbon in CH_4 forms four strong, and equivalent bonds, in tetrahedral directions. The explanation lies in the important notion of the *hybridization of a.o.* The 'mixing' of one s-a.o. with three p-a.o. leads to four equivalent sp^3-hybridized a.o., which are directed towards the corners of a regular tetrahedron. This can be roughly demonstrated in the following way. Omitting scaling-factors, the formal, independent combinations of a.o. can be represented by the expressions:

(i) $s + p_x + p_y + p_z$, (iii) $s - p_x - p_y + p_z$,

(ii) $s + p_x - p_y - p_z$, (iv) $s - p_x + p_y - p_z$.

Figure 5.4(a) signifies schematically the three mutually perpendicular p-a.o., each with its positive and negative lobes indicated by full and open circles. Combination of the three p-a.o., with the signs appropriate to (i), will yield a positive region in the direction suggested

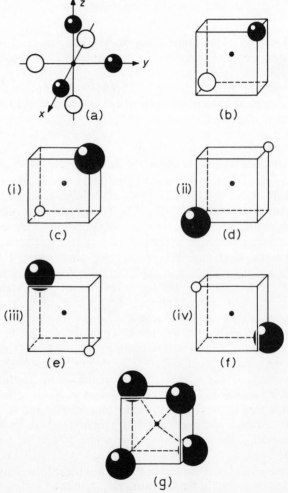

Fig. 5.4 Formation of four tetrahedrally directed orbitals by sp^3-hybridization.

by the full circle of Fig. 5.4(b), and a negative one in the opposite direction, shown by the open circle. When the combination is completed by addition of the s-a.o., supposed to be in its positive phase, the total positive density will be increased and the negative diminished, giving the overall effect symbolized by the full sphere in Fig. 5.4(c). Combination (ii) reverses the directions of p_y and p_z, so that, with s added positively, we get a positive accumulation in the region shown in Fig. 5.4(d); and similarly the other two combinations lead to accumulations as shown in (e) and (f). Taking all four hybridized orbitals, we have the arrangement shown in (g): the four are directed towards alternate corners of the cube—in mutually tetrahedral directions.

The highly directional character of sp^3-hybridized a.o. enables them to overlap effectively with the $1s$-a.o. of four H-atoms. The energy liberated in the formation of four strong σ-bonds compensates for the energy used in promotion and hybridization. The eight valency electrons are thus accommodated in four bonding m.o.

Hybridization is a useful concept, but the reader must not take it too literally. An atom of carbon, sensing an opportunity of combining with four H-atoms, does not actually go through a real process of promotion followed by hybridization. Hybridization in this sense does not occur. Rather it is a procedure at the service of the theoretician: if he puts the a.o. through a mathematically valid process of mixing, the mixed a.o. gives him a better approximation to the true wave-function of the CH_4-molecule than he could achieve with unhybridized orbitals.

5.6 Other forms of hybridization, the ethylenic bond

sp^3-Hybridization provides a convenient basis for the formal consideration of the bonding in saturated hydrocarbons and their derivatives. Other ways of mixing a.o. are possible, and many of them find application in valency theory. An important example is sp^2-hybridization, the mixing of one s- with two p-orbitals. This leads to three equivalent orbitals, lying in a plane and each at $120°$ to the other two. If we regard the unhybridized p-orbital as having its lobes directed towards the north and south poles, the hybridized orbitals lie in the equatorial plane, as suggested in Fig. 5.5(a). We may suppose a carbon atom to be in this initial condition before we seek to construct a molecule such as that of ethylene containing a double bond. As sketched in Fig. 5.5(b), overlap of these hybridized a.o. with $1s$-a.o. of H-atoms, or with the hybridized a.o. of the other C-atom, leads to m.o. corresponding to a C—C bond and four C—H bonds.

All are σ-bonds; in particular the C—C bonding is not affected if the two halves of the molecule are rotated about the bond-direction. In each of these m.o. we have accommodated two electrons. We still have two unplaced electrons, and each C-atom has its unhybridized p-orbital. Provided the mutual orientations are correct, the lobes of the p-orbitals will overlap, as is suggested in Fig. 5.5(b). This sideways overlap leads to two m.o., one bonding, the other antibonding; the former arises when the signs of the p-lobes are in phase as indicated. The resulting m.o. has a shape, roughly described as a 'double sausage', suggested in Fig. 5.5(c). There is a nodal plane passing through the six atoms. This m.o. accommodates the remaining two electrons, and thus sets up a second bond between the C-atoms. Orbitals of this sort, formed only when the component p-orbitals are in a suitable relative orientation, are known as π-orbitals.

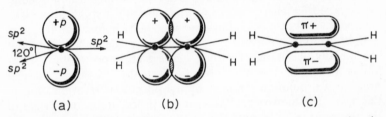

(a)   (b)   (c)

Fig. 5.5 sp^2-Hybridization of a carbon atom, and its use in interpreting the ethylenic double bond: formation of a π-molecular orbital by lateral overlap of two p-atomic orbitals.

From this point of view, the double bond is not just two single bonds. It is better described as a σ-bond plus a π-bond. The two halves of the molecule are strongly inhibited from rotating with respect to one another; for, if they turned through $90°$, the π-bonding would be lost, and the energy of the molecule would have to be raised by 50–100 kcal/mole.

The familiar double bonds of organic chemistry can be formulated in this way: e.g. C=O bonds in aldehydes or ketones or the C=N bonds in imides. However, double bonds of this type are probably limited to elements of the First Short Period. We sometimes use formulae involving double bonds between (say) an S-atom and an O-atom, as in §3.7. In such cases d-orbitals of the S-atom are probably involved, and the π-bonding is less simply represented than above.

Triple bonds can be interpreted by sp-hybridization: two a.o. are formed by mixing one $2s$- and one $2p$-orbitals of the C-atom. The hybridized a.o. are at $180°$ to one another. Thus in the acetylene molecule the H—C—C—H skeleton is formed by σ-bonding. Each C-atom still has two unhybridized p-orbitals at right-angles to each other and to the long axis of the molecule. Side-ways overlap leads to two bonding m.o. of π-character, and these take the remaining four electrons.

Two other types of hybridization are especially useful: dsp^2 gives four a.o. at $90°$ intervals directed towards the corners of a square; d^2sp^3 leads to six a.o. directed towards the corners of a regular octahedron. The latter may be supposed to occur in the SF_6-molecule.

5.7 Conjugated double bonds and aromatic systems

The m.o. we have considered so far have been formed between a pair of atoms, so that the electrons occupying such an m.o. are formally localized between the two atoms: the m.o. are two-centre orbitals. An extension of the notion of π-bonding leads us to m.o. which cover several atoms. Electrons supposed to be accommodated in such an m.o. are *delocalized*. The simplest example occurs in the benzene molecule—the prototype of all aromatic systems. We may formulate the benzenoid ring with six C-atoms in their sp^2-hybridized state. Overlap of these orbitals with each other or with $1s$-orbitals of H-atoms leads to twelve σ-bonds as suggested in Fig. 5.6(a). We still have six unused electrons and each C-atom has it unhybridized p-a.o. perpendicular to the plane of the ring. Lateral overlap of the sort represented, very schematically, in Fig. 5.6(b) leads to the formation of six m.o. Three of these are bonding; the simplest of them is sketched in (c). All three π-m.o. have a nodal plane through the nuclei of the twelve atoms, and they accommodate the six

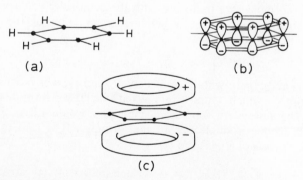

(a)   (b)

(c)

Fig. 5.6 Molecular orbital interpretation of the aromatic ring: (c) represents one π-type m.o., with a node in the plane of the ring. (This and two other, more complex, π-orbitals, accommodate the sextet of aromatic electrons.)

'aromatic electrons'. The conjugated system of double bonds in the molecule of $1:3$-butodiene can be treated similarly (§7.5).

With certain assumptions accepted, quantitative calculations can be done on the basis of this model of the aromatic system. They enable us to derive correlations between the measurable properties of different aromatic molecules.

5.8   The valency-bond approximation and resonance

According to the Heitler–London approach (see §5.3), the bonding energy of the H_2-molecule is mainly accounted for by approximating the actual wave-function by a combination of the two function, ϕ_1 and ϕ_2, which differ only in that the electrons are interchanged :

$$\phi_1, \; H(1)e(1)\,H(2)e(2); \quad \phi_2, \; H(1)e(2)\,H(2)e(1).$$

Though it would be difficult to do the corresponding calculation for a larger molecule, we may reasonably suppose that a similar explanation might hold for any covalent bond.

The same idea can be applied more widely. Just as H_2 acquires stability because it is, in a sense, a 'blend' of ϕ_1 and ϕ_2, so any molecule will acquire extra stability if we can write for it more than one electronic bond-diagram. We may suppose its electronic structure to be better approximated by a 'blend' than by either formula alone. For example, the most obvious electronic formula for the CO_2-molecule is (2). But we can also write (1) and (3)*.

$$|O\equiv C-\overline{\underline{O}}| \;\; (1) \qquad \langle\!\langle O=C=O \rangle\!\rangle \;\; (2) \qquad |\overline{\underline{O}}-C\equiv O| \;\; (3)$$

All three formulae are in accord with the octet rule. Valency-bond theory then supposes that a better representation of the CO_2-molecule is

$$\Phi(CO_2) \approx c_1\phi_1 + c_2\phi_2 + c_3\phi_3.$$

(Since (1) and (3) are wholly equivalent, c_1 and c_3, which define the proportions in which (1) and (3) are mixed with (2), will be numerically equal.) The molecule will be more stable than would a hypothetical molecule having a single one of the formulae. It is said to be a *resonance hybrid* of (1), (2) and (3).

---

* The reader should check that the O-atoms in (1) and (3) carry formal charges, which may be shown optionally by, for instance, writing (1) as

$$\overset{\oplus}{O}\equiv C-\overset{\ominus}{O}.$$

Resonance may be similarly invoked to explain the stability of an aromatic molecule such as that of benzene. Kekulé accounted for the equivalence of all positions round the benzene ring by writing formulae (4) and (5) (from which we have omitted the H-atoms for simplicity). He supposed that the double bonds were oscillating between the alternative positions shown in the two formulae. This explanation is taken over by v.b. theory, except that the bonds should no longer be regarded as oscillating. The molecule is not sometimes like (4) and at other times like (5): it is *all the time* a blend of the two.

(4) (5)

The resonance concept lends itself to picturesque analogies, which may appear helpful. Pauling likened a resonance hybrid to a mule, which is a cross between a horse and a donkey: not sometimes a horse, other times a donkey; but all the time a new sort of animal, which is a hybrid of the two. The weakness of this first-order analogy is that horses and donkeys really exist, whereas molecules with formulae (4) and (5) do not*. Molecules (4) and (5) are hypothetical affairs which we mix to obtain a better approximation to the real benzene molecule. A more exact, second-order analogy was therefore suggested by Wheland: a medieval traveller discovered the rhinoceros—an animal which really existed in remote parts of the world. To describe it to his friends at home, who had no concept of a rhinoceros, he said it was something between a dragon and a unicorn. He was using two mythological creatures to approximate to an unknown reality.

The difference between m.o. and v.b. approximations can be usefully emphasized here. Both methods employ an expression of the type, $\Psi = c_1\psi_1 + c_2\psi_2 + \ldots$. In m.o. theory the symbols ψ stand for orbitals. With the benzene molecule, for instance, six m.o. are constructed from the six p-a.o. of carbon; the six aromatic electrons are then supposed to be accommodated in the three bonding m.o.—the three of lowest energy. In v.b. theory the symbols ϕ stand for the whole molecule; the wave-function for the whole molecule is approximated in terms of wave-functions for simpler, hypothetical molecules.

* At least not in the substance we know as ordinary benzene.

Useful quantitative calculations can be done with both approximations, and the methods are complementary. To chemists—and certainly to those of the older school—v.b. theory seems more 'natural' because it makes use of the long-familiar notion of a localized bond between atoms. We have just used the idea of three extra bonds which can be shared between six positions round the benzenoid ring. The m.o. picture of six electrons occupying three non-localized orbitals arrives at much the same final result without mentioning bonds; the electrons are delocalized over the ring, and the energy of the whole molecule is lowered.

5.9 The valency-bond interpretation of molecular data

In Chapter 4 we saw how metrical information has become available on the sizes, shapes and other properties of molecules. Pauling used the concept of resonance to give a simple interpretation of some of this information. He laid down conditions which must hold before we can postulate resonance between two or more possible electronic bond-diagrams. The proposed forms must correspond to hypothetical molecules which would have (a) their atomic nuclei in approximately the same positions, (b) the same energy, or nearly so, and (c) the same number of unpaired electron spins. The more a proposed contributory form deviates from conditions (a) and (b), the less will its importance be: the smaller will be the coefficient representing its share in the 'mixture'.

When these conditions are adequately satisfied, the actual molecule is to be regarded as a hybrid of the various forms. Observable consequences are (i) that the atoms are rather closer together than would be expected by merely averaging the interatomic distances in the contributory forms, and (ii) that the energy of the actual molecule is rather less than would correspond to any single form.

We illustrate these principles by two examples. The covalent bond-radii in Table 4.4 enable us to estimate the C—O distances in the carbon dioxide molecule if it had any of the three formulae (1)–(3) of 5.8. In (2) each bond is double, and the calculated length is 1.21 Å. In either (1) or (3) the single bond would have a length of 1.43 and the triple bond 1.10 Å. The observed C—O distances are both the same, 1.16 Å, which is decidedly less than the double-bond value, and less than the average of 1.43, 1.21 and 1.10 Å. The shortening is interpreted as a symptom of the extra stability due to the resonance.

In Table 4.5 we listed some bond-energy terms, from which the energy of formation of a gaseous molecule—from its atoms—can be assessed. In a single Kekulé formula for benzene, we have six C—H bonds, three C—C and three C=C: adding the appropriate terms we find $6 \times 99 + 3 \times 83 + 3 \times 146 = 1281$ kcal/mole. The actual value derived from the observed heat of formation of benzene, with due allowance for the energy needed to atomize the elements, is about 1320 kcal/mole. Some 40 kcal more energy is liberated when a mole of actual benzene is formed than we should expect if the molecule had the single formula (4), or (5). This difference is the *resonance energy*, which is taken as a measure of the extra stability attributable to resonance between (4) and (5).

One application of the resonance concept is embodied in Fig. 5.7, which plots the length of a C—C bond against the *bond order*. In the simplest possible approach we take the C—C distance of 1·54 Å in ethane as appropriate to the pure single bond (of order 1); that in ethylene (1·33 Å) for the double bond (order 2); and that in benzene (1·39 Å) for a bond of order 1·5. This last is based on the v.b. idea that the molecule is a mixture of the two Kekulé forms—(4) and (5), necessarily in equal proportions—so that each bond is half-way between single and double.

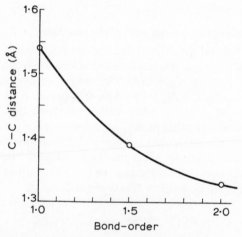

Fig. 5.7   Rough correlation-curve between bond-order and C—C distance.

Having established such a curve, we can use it to assess the order of some other C—C bond whose length we have determined. For example, the 1:2 and 2:3 bonds in the naphthalene molecule differ slightly in length according to an accurate X-ray study: the C—C distances are, respectively, 1·36 and 1·42 Å. The curve in Fig. 5.7

suggests that the order of these bonds should be about 1·7 and 1·4. Making the assumption that this molecule can be regarded as a resonance hybrid to which the forms (6), (7) and (8) contribute equally, the reader should have no difficulty in convincing himself that the 1:2 and 2:3 bonds have orders of 5/3 and 4/3. In this particular case, the agreement is better than could have been expected from the simple-minded assumptions we have made and from the uncertainty of drawing a curve through three fixed points.

More sophisticated calculations of bond-order can be made by both a.o. and v.b. methods. By making use of several different molecules, a number of points can be plotted in an attempt to establish a bond-order v. bond-length curve; but there is always some 'scatter' of the points, and, in any case, the experimental bond-lengths are liable to uncertainties no less than the theoretical estimates of order. A similar treatment can be applied to bonds between other types of atoms.

(6)   (7)   (8)

5.10   Odd molecules—a resonance interpretation

In §3.8 we stated the evident fact that no electronic bond-diagram, satisfying the simple rules, can be written for a molecule such as NO with an odd number of electrons. One way of explaining the relative stability of such molecules is based on v.b. theory. We may write two formulae for NO, (9) and (10), neither of which is satisfactory by itself because one or other of the atoms has only a septet of electrons. We may then postulate that resonance between these forms causes sufficient stabilization to offset the inadequacy of either alone. The reader should apply the same idea to the NO_2-molecule, for which four formulae can be written, (11) being one.

(9)   (10)   (11)

Though not 'odd' in the literal sense, the O_2-molecule has properties which preclude the obvious electronic formula (12).

(12)   (13)   (14)

Oxygen is, for instance, paramagnetic, which is a sign of unpaired electron spins. The v.b. interpretation is to take the molecule as a resonance hybrid between (13) and (14). Presumably there is a delicate balance of considerations which render a paramagnetic molecule—approximated as a blend of (13) and (14)—rather more stable than (12). This is certainly a significant fact for chemistry and biochemistry; the reactivity of oxygen is of great importance, whilst a substance whose molecules had formula (12) would probably be as stable as nitrogen.

5.11 Inert-gas compounds and resonance

A v.b. interpretation can be offered for the relative stability of such molecules as XeF_2. We may first note that this molecule is isoelectronic with the hypothetical ion IF_2^-. The number of electrons is the same, but the lower nuclear charge of I results in the net negative charge. Though IF_2^- is not yet known, the analogous ICl_2^- is well established, and has a linear structure like that of XeF_2. The following electronic transaction is in accord with the elementary rules:

$$|\overline{\underline{Xe}}| + 2|\overline{\underline{F}} \cdot \longrightarrow |\overline{\underline{F}} - \overset{\oplus}{\underline{Xe}}| + |\overset{\ominus}{\underline{\overline{F}}}|.$$

As it stands the right-hand side is not a valid representation, since xenon difluoride is a molecular, and not an electrovalent, compound. However, Coulson suggests that the best, simple way of regarding this molecule is to think of it as a resonance hybrid between two forms (15) and (16). The bonds are thus half covalent and half ionic.

$$|\overline{\underline{F}} - \overset{\oplus}{\underline{Xe}}| \ \overset{\ominus}{|\underline{\overline{F}}|} \qquad \overset{\ominus}{|\underline{\overline{F}}|} \ |\overset{\oplus}{\underline{Xe}} - \overline{\underline{F}}|$$

$$(15) \qquad\qquad (16)$$

Chapter 6

The Hydrogen Bond and Other Forces

6.1 Electrovalency and covalency

In Chapters 3 and 5 we presented electrovalency and covalency as quite distinct sorts of interatomic force. The one is directly electrostatic; it arises between ions formed after one or more electrons have been transferred from one atom to another. The other is attributed to a sharing of pairs of electrons, whereby, via a wave-mechanical effect, electron density accumulates between the atomic cores. The latter effect is indeed electrostatic too, but the mechanism is different in kind.

Nevertheless the current view is that there is no hard-and-fast distinction. Whilst many bonds, e.g. in Na^+Cl^-, are preponderantly ionic, and many others, e.g. in C_2H_6, are preponderantly covalent, intermediate types do occur. We can discuss this idea most simply in v.b. terminology. The original Heitler–London treatment of the H_2-molecule supposed it to be approximately represented as a blend of (1) and (2).

$$H(1)e(1) \quad H(2)e(2) \qquad\qquad H(1)e(2) \quad H(2)e(1)$$
$$(1) \qquad\qquad\qquad (2)$$

We may improve our approximation by adding two other forms, (3) and (4):

$$\overset{\ominus}{H}(1)e(1)e(2) \quad \overset{\oplus}{H}(2), \qquad\qquad \overset{\oplus}{H}(1) \quad \overset{\ominus}{H}(2)e(1)e(2).$$
$$(3) \qquad\qquad\qquad\qquad (4)$$

The approximate wave-function would then be

$$c_1\phi_1 + c_2\phi_2 + c_3\phi_3 + c_4\phi_4;$$

and, because of the identity of (1) and (2) and of (3) and (4), c_1 and c_2 must be numerically equal; so must c_3 and c_4, though we should expect them to be much smaller than c_1 or c_2.

Next we consider a case where the atoms differ in type, HCl:

$$He(1) \quad Cl\,e(2) \qquad He(2) \quad Cl\,e(1) \qquad \overset{\oplus}{H}\ \overset{\ominus}{Cl}\,e(1)e(2) \qquad \overset{\ominus}{H}e(1)e(2)\ \overset{\oplus}{Cl}.$$
$$(5) \qquad\qquad (6) \qquad\qquad (7) \qquad\qquad (8)$$

Now $\Phi(HCl) \approx c_5\phi_5 + c_6\phi_6 + c_7\phi_7 + c_8\phi_8$; c_5 and c_6 must be equal in magnitude. But c_7 and c_8 will differ, since, owing to the greater electronegativity of chlorine, (7) will be much more stable than (8). Hence (8) can be neglected and we are left with

$$\Phi(HCl) \approx c_5\phi_5 + c_6\phi_6 + c_7\phi_7.$$

In v.b. terms, the HCl-molecule is a resonance hybrid between a hypothetical, purely covalent form—itself a blend of (5) and (6)—and a hypothetical, purely electrovalent form (7). The bond is to be regarded as mainly covalent, though with some admixture of *ionic character*.

One very rough way of assessing the amount of ionic character depends on the dipole moment. If HCl were purely electrovalent, there would be full positive and negative charges (each amounting to $e = 4.8 \times 10^{-10}$ e.s. unit) on the respective atoms. As we know the interatomic distance (1.275 Å, see Table 4.1), we can easily calculate the moment for the H^+Cl^- molecule:

$$\mu(calc) = 4.8 \times 10^{-10} \times 1.275 \times 10^{-8} = 6.1 \text{ D}.$$

In fact $\mu(obs)$ is 1.07 D. Therefore we may suppose the bond to have $1.07/6.1 = 17\%$ ionic character. Since the dipole moment associated with a covalent bond can be loosely regarded as arising from the difference between the electronegativities of the bonded atoms, this difference is a rough measure of the ionic character. From this over-simplified point-of-view, most bonds are either mainly covalent or mainly electrovalent. (Cases such as the HF-molecule, whose bond has about 50% ionic character, are uncommon.) Therefore the distinction between electrovalency and covalency remains a useful one.

6.2   The metallic bond

We must now describe a number of other types of force that can operate between atoms and molecules. Most of them are relatively feeble, though they sometimes qualify to be thought of as valency-forces.

One of these forces however may be very strong. This is the *metallic bond* between the atoms of a metal. Because it does not easily fit into the electrovalency–covalency scheme, and because it gives rise to intermetallic compounds with formulae, such as $Cu_{31}Sn_8$, that do not even remotely correspond with conventional valency-numbers, the metallic bond has been neglected by chemists; though its practical importance is obvious. That the atoms of a metal such as nickel are held together by forces at least as strong as those in sodium chloride is evidenced by the mechanical strength, and by the high melting point and latent heat of vaporization. Most metals have exceptionally high electrical conductances. From an electronic point of view metals are elements with, in their outermost shells, a small number of electrons that can be fairly easily removed. Iron, for instance, with the structure $1s^2\, 2s^2\, 2p^6\, 3s^2\, 3p^6\, 3d^6\, 4s^2$, easily loses two or three electrons.

The situation in a metal can be treated by both m.o. and v.b. theory. The general picture is one of a crystalline array of metallic ions held together by a matrix of relatively free electrons. The valency electrons are delocalized over the whole crystal.

6.3   Other forces

We now list and summarize some weaker interactions between atoms and molecules.

(i) *Ion–dipole force.* When a polar molecule is near to an ion, it will tend to become orientated in the way suggested by Fig. 6.1(a). The molecule may be revolving, but it will then spend more of its time in the direction shown than in any other. The potential energy will be minimized in this position, since the attraction between oppositely charged points will exceed the repulsion between the more distant like charges. An overall attraction of moderate strength will result.

This force account for at least much of the *hydration of ions* which occurs when a salt is dissolved in water*. The ions have an intense electrostatic field, and they attract the dipolar water-molecules. The hydration is particularly effective with a positively charged ion; a number of water molecules can be arranged neatly round a cation in a way suggested in Fig. 6.1(b). (Coordination by some form of dative bonding may also play a part: see §3.5.) The water molecules in direct contact with the ion will be 'permanently' held: a relatively long time may elapse before they are replaced by other molecules, and they will move bodily with the ion. Beyond the limits of this inner hydration sheath, other water molecules may be more loosely bound. Because of the diminishing force as we move further away from the ion, attempts to estimate the hydration numbers of particular ions have led to very different values. It depends what we mean by 'hydration'. Perhaps four or six water molecules may be in the inner sheath; but as many as thirty or fifty may be under some sort of constraint by the electrostatic field of the ion.

---

* And which is a dominant factor in making water a good solvent for electrolytes.

(*ii*) *Ion-induced dipole force.* When the molecule is not polar, the first effect cannot operate. However, when any molecule finds itself near to an ion, the intense field will tend to polarize it: a positive ion will repel the atomic nuclei and attract the electrons of the molecule; and so an *induced dipole* moment will result so long as the ion is near enough. Such an induced dipole will necessarily be in the sense depicted in Fig. 6.1(a), and an attractive force will operate, though a weak one. We note that this second effect will exist whether the molecule is polar or not; if the molecule already has a permanent moment, it will add its lesser contribution to that already present.

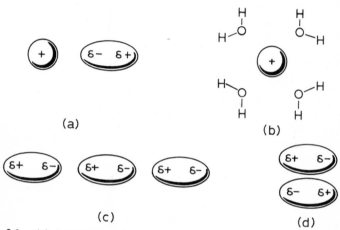

Fig. 6.1 (a) Ion-dipole attraction; (b) possible arrangement of polar H_2O-molecules round a cation; (c) and (d) dipole–dipole attraction.

(*iii*) *Dipole–dipole force.* Neighbouring polar molecules will tend to orientate themselves so as to minimize the energy in one or other of the styles sketched in Fig. 6.1(c) and (d). This will lead to a weak net attraction.

(*iv*) *Dipole-induced dipole force.* The name of this very weak force probably suffices to explain matters. A polar molecule, when near to another molecule—polar or not—will induce in it a temporary dipole.

(*v*) *The dispersion force.* Though they are normally too weak to give compound formation, forces (iii) and (iv) reveal themselves in a general attraction between neutral molecules. One consequence is the deviation of gases and vapours from ideal behaviour required by the equation $PV = RT$. Part of this deviation is recognized, for instance, in the a/V^2 term of van der Waal's equation of state, $(P + a/V^2) \times$

$(V - b) = RT$. A second, and related, consequence is that all gases, if cooled and compressed sufficiently, condense to liquids—a state of matter in which cohesive, intermolecular forces are dominant. Since these consequences apply to all substances, including those whose molecules have zero dipole moments, and even the inert gases, it follows that there must be some other type of intermolecular force besides those listed hitherto. That such a force could arise by a wave-mechanical effect was shown by London. Since the electrons of any molecule are in motion, temporary dipole moments may be supposed to arise, when the centres of action of positive and negative charges in the molecule are, transiently, different. Thus the temporary dipole in molecule A will induce a dipole in an adjacent molecule B. So also the motions of electrons in B will produce temporary dipole moments which will induce a moment in A. It turns out that these respective effects tend to get into phase with one another, and thus that an attractive force develops. This *dispersion force* is weak, but it must operate between any pair of neighbouring molecules. Together with (iii) and (iv) where they apply, it constitutes the *van der Waals force* of attraction, so-called because of its correspondence to the a-term in the equation of state.

(*vi*) *Repulsive forces.* All the forces we have discussed so far have been attractive. There must also be some repulsive force; otherwise molecules would collapse. It is a characteristic of such repulsive forces that they increase very suddenly with the approach of two atoms or molecules: the force is negligible beyond a certain distance, but at that critical distance it starts to rise rapidly. Perhaps it follows an inverse-power law of the type, repulsion $\propto (1/d^n)$, where n is a high power of the distance, d. Not much is known about this force, but presumably it is connected with the reluctance of the electrons of different entities to be forced too closely together. It is the repulsive force that gives the atom a rather well-defined 'boundary' at the van der Waals radius (see §4.16).

6.4 The hydrogen bond

Water has long been recognized to be a most unusual compound. To cite a few examples, it remains a liquid up to 100°, under ordinary conditions, though its molecular weight is only 18, whereas most compounds with double this weight are gases. It expands on solidification, a very rare phenomenon. It dissolves salts, a thing done by few liquids. It has a very high specific heat.

Similar, though less marked, anomalies characterize compounds whose molecules contain hydroxyl groups. Ethanol, C_2H_5OH, boils at 78°; propane, C_3H_8, with a nearly equal molecular weight boils at $-42°$. The same often holds for compounds involving amino-groups: ethylamine, $C_2H_5NH_2$, boils at 17°. These anomalies disappear when the H-atoms of the OH- or NH_2-group are replaced by an alkyl radical (R). Diethyl ether, R_2O (where $R = C_2H_5$), though its molecular weight is almost double that of ethanol, is a more normal liquid, boiling at 35°.

Long ago chemists realized that these anomalies were the result of molecular association. Liquid water, they felt, consisted of polymeric molecules, $(H_2O)_n$, rather than of single H_2O-molecules. The extent of association was unknown; indirect inference suggested that n might be 2 or 3. How the H_2O, or ROH, molecule, which seemed to be saturated in the valency sense, contrived to combine with other similar molecules was unknown. The fact that anomalous behaviour disappeared when ROH became ROR suggested that polymerization somehow depended on the H-atom.

The logic of this is that the H-atom must be able to link two other atoms together. In other words, hydrogen must in some circumstances become quasi-bivalent. Since the univalency of hydrogen had been held as a basic doctrine of chemistry, this conclusion was resisted. However by 1908 percipient chemists began to write formulae embodying this idea. Werner, for instance, represented the anion of the salt KHF_2 as $F—H---F^-$. In 1912 Moore and Winmill explained the fact that the bases NH_3, $NH_2(CH_3)$, $NH(CH_3)_2$ and $N(CH_3)_3$ are weak, whilst the quaternary base $N(CH_3)_4OH$ is very strong, in the following way. When dissolved in water, the first four bases all combine with the solvent to yield unionized forms such as (1), which are only slightly ionized. The equilibrium is far to the left in the equation shown. On the other hand the tetramethyl ammonium hydroxide, lacking a H-atom on the nitrogen, cannot exist in a form analogous to (1); hence in its case the compound exists only in the completely ionized form (2), and so is a strong alkali.

The association of an anomalous liquid such as water could be represented similarly by such a scheme as (3).

$$
\begin{array}{c}
\text{H} \\
| \\
\text{H—O—H- - -O} \quad \text{H} \\
\qquad\qquad | \qquad | \\
\qquad\quad \text{H- - -O} \\
\qquad\qquad\quad | \\
\qquad\qquad\quad \text{H}
\end{array}
$$

(3)

Molecules so linked are now said to be joined by a *hydrogen bond*. Though the bonding is rather weak, its consequences are important in many directions besides that of molecular association.

6.5   The properties of the hydrogen bond

The hydrogen bond is formed in the situation X—H- - -Y, where X and Y are atoms, or groups, of sufficient electronegativity*. Strong bonding is confined to the cases where X and Y are the elements F, O and N; but weaker bonding may occur in otherwise favourable circumstances with Cl, S and even C (when this is bonded to other more electronegative atoms).

The presence of hydrogen bonding is often implied by an experimental determination of the distance $X \ldots Y$. For illustration we shall take the case where X and Y are both oxygen. Were these atoms directly bonded by covalency, the distance between them would be less than 1·4 Å. If, on the other hand, there was no bond at all, the distance should exceed the sum of the van der Waals radii (2 × 1·4 = 2·8 Å; see §**4.16**). If the O ... O distance lies between these limits, a hydrogen bond is to be suspected, provided—as is always the case—there are chemical grounds for knowing a H-atom to be available, attached to one of the O-atoms. In fact, this consideration makes it necessary to raise the upper limit beyond 2·8 Å. The O—H distance is around 1·0 Å; add to this the van der Waals radius of the other O-atom and that of the H-atom itself, which is around 0·9 Å, and we arrive at 3·3 Å. When the observed O ... O distance is less than this, and when there are stereochemical reasons for believing that a H-atom lies on, or near, the O ... O-line, then a hydrogen bond is probably present.

* From another point of view, X—H must be sufficiently acidic and Y sufficiently basic.

Crystal-structure analysis is the most suitable way of detecting hydrogen bonding by this criterion. We find O...(H)...O distances ranging from the upper limit ($\sim 3 \cdot 3$) down to 2.4 Å. In ice, for instance, each O-atom has four nearest O-neighbours at a distance of $2 \cdot 76$ Å, corresponding to rather weak bonding (see Fig. 6.2). In a carboxylic acid dimer (4), O...O is usually about $2 \cdot 65$ Å—a moderately strong bond. In some of the acid salts, formed rather surprisingly by many monocarboxylic acids, such as sodium hydrogen diacetate ($NaH(CH_3CO_2)_2$) we find distances down to $2 \cdot 45$ Å or rather less.

$$R-C \underset{O---H-O}{\overset{O-H---O}{\big\langle}} C-R$$

(4)

These are very strong hydrogen bonds. Hydrogen bonds of the type O—H---N are usually weaker, with O...N distances in the range $3 \cdot 3$–$2 \cdot 8$ Å. The F—H---$\overset{\ominus}{F}$ bond is the shortest and strongest hydrogen bond; in the bifluoride ion, HF_2^-, the F...F-distance is $2 \cdot 26$ Å. The energy of the hydrogen bond ranges widely in a similar way to the length. The energy needed to dissociate the strong bond in HF_2^- is perhaps 40 kcal/mole—an exceptionally high value. For the O—H---O-bonds, the range is from about 12 kcal down to 1–2 kcal for the longest, and weakest, bonds. In general hydrogen bonds are about ten times weaker than typical covalent bonds (see Table 4.5).

Directional considerations are important. Reasonably strong hydrogen bonds can be expected only when the proton can lie on, or near to, the O...O-line. In the examples given above this is possible. In a molecule such as that of catechol (5) it is stereochemically impossible for either of the hydroxylic H-atoms to be near the O...O-line, as the normal C—O—H angle is around 105°. Though the two O-atoms are forced by the bond-structure of this molecule

(5)

to be only about 2·8 Å apart, in this case the discovery that such a distance actually obtains is not good evidence of hydrogen bonding.

The commonest method for studying crystal structures is by X-ray diffraction. This method is not very efficient for locating H-atoms directly. So, in all the older work, as in much of the more recent, the position of the proton in an otherwise well-documented hydrogen bond can be inferred only indirectly. The evidence indicates that the proton is normally much closer to one of the electronegative atoms than to the other. The situation implied by O—H- - -O usually obtains. This conclusion has been largely based on a study of the infra-red spectra of hydrogen-bonded compounds. An isolated OH-group gives rise to a strong absorption band in the region of 3500 cm^{-1} ($\lambda = 28,000$ Å). This absorption is associated with the vibrational stretching of the O—H-bond. In most hydrogen-bonded compounds the corresponding absorption is broader and shifted to lower frequency (~ 3000–2500 cm^{-1}). The relevant point is that the frequency, though changed, is not greatly changed; the proton vibrates as though it is still basically attached to one O-atom, rather than suspended between two.

More direct evidence comes from neutron diffraction by crystals. In a number of crystals, X-ray analysis has now been followed by neutron diffraction, which enables us to locate the proton much more accurately. In nearly all cases the proton has been found to be definitely attached more closely to one of its neighbours. For example, Robertson's X-ray analysis of resorcinol, m-$C_6H_4(OH)_2$, showed the molecules to be linked together by hydrogen bonds into a complicated set of interconnected chains. The O ... O-distance was about 2·72 Å—a weak hydrogen bond. Neutron-diffraction analysis by Bacon and Curry (1957) located the protons 1·02 ± 0·04 Å from one O-atom and hence 1·70 Å from the other. Most of the compounds studied prove to be of this type.

However, from the limited number of crystals for which reliable neutron-diffraction studies have been made it appears that, as the O ... O-distance diminishes (i.e. as the bond becomes stronger), the O—H-distance increases. This seems reasonable, as the approach of the second O-atom will tend to slacken the original O—H bond. If this effect continues to hold, and if the overall length of the bond becomes short enough, the proton would ultimately find itself at the centre. This consideration adds interest to the study of very short hydrogen bonds with O ... O less than 2·45 Å, and of the unusually strong bond in the HF_2^--ion. We shall revert to these exceptional cases in the next section.

6.6 The nature of hydrogen bonding

In a hydrogen bond the H-atom is apparently bivalent: X—H - - - Y. The second bond cannot be covalent (as in X—H - - - Y), since the $1s$-orbital of the H-atom could not accommodate four electrons. (Were the extra electrons in the $2s$-orbital, their energy would be too high to give any effective bonding.) Two types of mechanism have been proposed.

The first, and older, attributed the bonding to resonance. In valency bond terms, we can write two electronic formulae:

$$X—H: Y \quad (6) \qquad \text{and} \qquad X: H\text{———}Y \quad (7);$$

and we may suppose the true structure to be a hybrid of these two. The binding (a few kcal/mole) was regarded as resonance energy. The objection to this lies in the fact that the proton is usually much closer to X than to Y. Hence the form with the unnaturally stretched bond (as we have shown it in (7)) between H and Y, and the unbonded X and H, crushed uncomfortably together, would have a much higher energy than (6). Resonance thrives only when the contributory forms are of comparable energy. This mechanism cannot account for much of the bonding energy in most hydrogen bonds.

The second mechanism is an electrostatic one—closely related to the ion–dipole and dipole–dipole forces described in **§6.3**. Hydrogen bonding occurs when X and Y are electronegative. The X—H bond will therefore have a moment in the sense shown below; and Y will also carry some negative charge—a partial charge if it is attached to some other, less electronegative, atom; a full charge if it is an anion (as in HF_2^-). In the orientation shown a net attractive force will result,

$$\overset{\delta-}{X}\text{—}\overset{\delta+}{H} \qquad \overset{\ominus \text{ or } \delta-}{Y}$$

since the unlike, attracting charges are closer together than are the like, repulsive charges. Fairly simple calculations can be based on this model, and they suggest that the electrostatic effect is adequate to account for the binding energy in the case of weak, or moderately strong, bonds.

The evidence thus far presented is wholly in favour of the electrostatic against the resonance model. This may not be so certain for the less common class of very strong bonds—that in HF_2^-, or the O—H—O bonds whose lengths fall below 2·5 Å. In the first place the infrared spectra of compounds containing these bonds are anomalous; the O—H, or F—H, stretching frequency is greatly reduced, and the width of the absorption band greatly increased—so much so that the band may no longer be recognizable. We conclude that the unperturbed O—H bond no longer survives in very strong hydrogen bonds. Secondly, as we stated in §6.5, the proton moves towards the mid-point as O...O diminishes. If it reaches the centre, then the alternative bonding schemes, which are unequal in long bonds, become equivalent:

$$O-H: O \qquad\qquad O: H-O;$$

so that resonance stabilization would become important. (A corresponding m.o. interpretation is easily envisaged.)

Current opinion can probably be summarized as follows: in the majority of hydrogen bonds, the O—H bond retains its identity, and the electrostatic mechanism is substantially correct; but in very short bonds the proton tends to be displaced towards the centre and, as this happens, the resonance contribution becomes significant. Whether there are any hydrogen bonds with the proton literally at the mid-point is not certain—though it is nearly certain that HF_2^- is genuinely symmetrical; but the very short O—H—O bonds are possible examples of symmetrical hydrogen bonds, and they are likely to be eagerly studied as more sensitive experimental methods become available for locating the proton.

6.7   The rôle of hydrogen bonding

Though rather weak, the hydrogen bond is important. When co-valent molecules form a crystal, there are often various alternative ways in which they could pack themselves together. In such cases the hydrogen-bond energy, though modest in amount, will confer some benefit on a mode of packing which allows the maximum of hydrogen bonding between neighbouring molecules. We find that this nearly always happens. Molecules are 'clever' at arranging themselves so as to form as many hydrogen bonds as possible. The sugars are examples. Their crystal structures are such that all, or most, of the OH groups are able to engage in hydrogen bonding, which is connected with the fact that many sugars crystallize well. On the other hand, the sugars play an important rôle in biological systems because their molecules are able to form hydrogen bonds with water. This accounts for their being much more soluble in water than one might expect for moderately large organic molecules: contrast glucose, $C_6H_6(OH)_6$ with cyclohexane, $C_6H_6H_6$. Protein molecules consist of long chains of amino-acid residues connected by the peptide linkage. Such chains could adopt innumerable alternative conformations. The possibility of hydrogen bonding between the

NH-group of an amino-acid residue and the C=O-group of another residue, four places along the chain, seems to stabilize the helical structure which occurs in some sections of protein molecules, and which is so important in their biological functioning. According to the Crick–Watson hypothesis, hydrogen bonding plays a vital part in the genetic coding in the vast molecules of such materials as DNA.

Ice is a notable example of a hydrogen-bonded solid. Each O-atom is joined by rather weak hydrogen bonds (O . . . O = 2·76 Å) to four others in directions towards the corners of a (nearly) regular tetrahedron, as is sketched in Fig. 6.2(a). This confers an open structure on ice, and accounts for the unusual feature that the solid is less

(a)   (b)

Fig. 6.2   (a) Arrangement of O-atoms in ice. The broken lines represent hydrogen bonds, but the H-atoms are not shown. (b) Disordered arrangement of H-atoms along the O–––O direction in ice, as revealed by neutron diffraction.

dense than the liquid. With so long a hydrogen bond, there can be no question of the proton's being at the mid-point of the bond; and at first there was a problem of reconciling the four-fold symmetry of the O-atom's environment with the contrary requirement that only two H-atoms can be covalently bonded to any one O-atom. The full explanation was derived from neutron-diffraction work (1957), which supported a suggestion made by Pauling as early as 1932. This work showed 'half-hydrogens' lying at two points along each O . . . O-bond, as is suggested in Fig. 6.2(b). This of course cannot be taken literally. Each bond has only one, whole H-atom; but, statistically, over the whole crystal, there is a 50:50 chance of finding it at either of two possible sites. This randomness accounts for the anomalously high entropy of ice.

Chapter 7

Quantivalence, the Molecule and the Bond–Concept

7.1 The significance of valency as a number

The word, 'valency', is used in two different senses. The meaning implied in the title of this book, and generally in the text—though not always, is the *force*, or forces, by means of which atoms are joined together to form molecules or other stable aggregates. The word is then being used qualitatively to describe the power atoms possess of combining chemically. The other meaning—and it is an older one— is quantitative; it is the *number of units* of such power appropriate to any atom. The carbon atom in methane is united to four hydrogen atoms, for example, and is said to be quadrivalent. Its valency is four. The word *quantivalence* was used by Roscoe and by Helmholtz; and in this chapter we may perhaps be permitted to use it when we need to refer specifically to valency in the numerical sense.

Methane consists of discrete molecules, CH_4, with the carbon closely joined to the four hydrogens; and the allocation of quantivalencies of four to the carbon and one to each hydrogen presents no difficulty. But the concept is not always so clear. In solid sodium chloride each sodium is 'joined' to six chlorines, each chlorine to six sodiums. Yet it would be unsatisfactory to conclude that each atom had a quantivalency of six; and still more unsatisfactory when the quantivalency in the similar compound, caesium chloride, proved to be eight. Then what is the quantivalency of the cobalt atom in $[Co(NH_3)_6]Cl_3$? Three, or six, or nine? And is that of the oxygen atom in ice two, or four? As we shall see later in this chapter, very much more difficult cases might have been posed.

A possible link between the two meanings of 'valency' can be found in the important notion of the *saturation of valency*. When two hydrogen atoms unite to give the molecule, H_2, we may suppose them to do so because each possesses an unpaired electron; in the molecule the spins are paired and the combining powers of the atoms are mutually satisfied. A third atom cannot be accepted by the system. This mechanism of saturation can be envisaged in all molecules involving straightforward covalency. Where electrovalency is concerned, the mechanism of saturation is quite different. Here it depends on the principle that the system must be electrically neutral. If a crystal of calcium fluoride contains N calcium ions, Ca^{2+}, it must necessarily contain exactly $2N$ fluoride ions, F^-, or very nearly so. The question of how many fluorides are directly in contact with each calcium is unimportant from this point of view.

7.2 Limitations of the concept of the molecule

These saturation mechanisms for covalency and electrovalency are so different that it is surprising that the idea of quantivalence was so successful in the latter half of the last century, when the distinction between the two kinds of force was not understood. The success can probably be attributed to the fact that valency theory developed largely in relation to organic, or simple gaseous inorganic, compounds, which consist of covalent molecules. Had the early inorganic chemists been in a position to study the structure of the solid state, then the progress of the theory might well have been slower.... So long as sodium chloride was thought to consist of Na—Cl molecules, it was appropriate to apply to it the same rules for determining the quantivalencies of the atoms as could be applied to methane. (Indeed we still imply this view when we speak of a 0·58 % solution of sodium chloride as 0·1M.) With the advent of physical methods of structure analysis, this Na—Cl molecule began to fade away—to the dismay of some chemists. It has been stated that, when the early X-ray crystallographers first announced that rock-salt was composed of sodium atoms (or ions) each surrounded by six equidistant chlorines, and vice versa, an eminent chemist implored them to look again until they found the Na—Cl molecule which must be there!

Quite apart from the presence of electrovalency, difficulties are apt to arise when we are dealing with the 'infinite assemblages' that occur in most crystals. A molecular formula must then be treated with suspicion. There is a deceptive similarity between the formulae, CO_2 and SiO_2; but, whilst the one substance consists of discrete triatomic molecules, the other is an infinite assemblage of SiO_4 tetrahedra, the oxygen atoms at the four corners of each being shared with neighbouring tetrahedra. Each silicon atom is covalently attached to four oxygens, each oxygen to two silicons, and so on throughout the crystal. No SiO_2 molecule can be separately distinguished in the structure. The formula, SiO_2, is valid only as representing the total composition. Even when we restrict ourselves to crystalline solids, such formulae can mislead. As Wells has pointed out, the compounds $CaCO_3$ and $CaTiO_3$ are wholly different in nature; the former is a salt composed of Ca^{2+} and CO_3^{2-} ions, the latter an oxide (perovskite structure), in which each calcium atom is coordinated to twelve oxygens, and each titanium to six.

There is a solid compound with the empirical formula $(NH_4)_2SbBr_6$, which suggests that the Sb-atom has an ionic quantivalence of 4, according to the schematic statement $Sb^{4+} + 6Br^- \longrightarrow [SbBr_6]^{2-}$. In terms which we shall explain in the next section, its oxidation state appears to be $+4$, or Sb(IV). In fact structure analysis suggests that there are two kinds of complex anions present: $[SbBr_6]^-$, and $[SbBr_6]^{3-}$, corresponding to Sb(V) and Sb(III) respectively. Crystals with such a duality of quantivalence are usually coloured. A substance which has attracted much interest is phosphorus pentachloride, PCl_5. The vapour does indeed contain PCl_5 molecules, though they tend to dissociate into PCl_3 and Cl_2. The molecular structure is one with the phosphorus atom at the centre of a trigonal bipyramid (4.9). In the solid state, however, the structure is astonishingly changed. X-ray analysis leaves little room for doubt that it is an ionic crystal composed of the complex ions, $[PCl_4]^+$ and $[PCl_6]^-$.

7.3   The oxidation state

An alternative to quantivalence in this sense is the *valence number*, or *oxidation number*, or *oxidation state*, introduced by Pauling. From the electronic point of view oxidation corresponds to the removal of electrons. The oxidation state therefore is the number of electrons lost by the elementary atom in attaining the condition it has in a given compound. Reduction—electrons gained—is represented by a negative oxidation state.

The oxidation state is defined as the formal charge on the atom after the electrons of the molecule have been allocated amongst the atoms according to the following rules. For each electrovalency the electrons are treated in the normal way; for each covalency both the shared electrons are supposed to be held by the more electronegative of the two bonded atoms but, when the atoms are of the same element, the pair is split between them. Thus, in sodium chloride, the oxidation numbers of Na and Cl are $+1$ and -1; in chloroform H $+1$, C $+2$, and Cl -1; in ammonium chloride H $+1$, N -3, Cl -1; in tri-methylamine oxide, $(CH_3)_3NO$, H $+1$, C -2, N -1 and O -2; in molecular nitrogen, each atom has zero oxidation numbers. The total of the oxidation states of all the atoms in a neutral molecule must necessarily be zero. For a charged molecule, the sum is equal to the positive ionic charge. Thus, in $[Co(NH_3)_6]^{3+}$, the oxidation states of Co, N and H are respectively $+3$, -3 and $+1$; and

$+3 - 6 \times 3 + 18 \times 1 = +3$. The oxidation states of metal atoms are commonly denoted by roman numerals, such as Co(III).

7.4   Crystal imperfections

There is another, and commoner, type of complexity liable to affect our judgment of quantivalence in ionic crystals. In any crystalline substance the *motif* is the unit cell; and in an *ideal crystal* this *motif* is exactly repeated throughout the region of space occupied by the whole crystal. In rock-salt, for example, the unit cell is a cube of edge 5·63 Å (and containing two 'molecules' of NaCl); so that an ideal crystal of external dimensions $0·1 \times 0·1 \times 0·1 \ cm^3$ would extend over $0·1 \div (5·63 \times 10^{-8}) = 1·77 \times 10^6$, unit cells each way, and would contain altogether $(1·77 \times 10^6)^3 = 5·6 \times 10^{18}$, unit cells, each exactly repeating the first. In fact such strict regularity hardly ever occurs. In nearly all crystals this perfection is marred by various kinds of defects. For instance, regularity of alignment persists only over relatively small distances—perhaps for 100 or 1000 unit cells in any one direction; then there will be some interruption before we come to another limited 'domain' of strict alignment. Such a crystal is said to possess a *mosaic structure*. Again it often happens that a site in the crystal which should be occupied by a particular kind of entity —a sodium ion perhaps—is actually vacant. Or alternatively an ion may not occupy a regular lattice position at all, but rather an *interstitial* position between the regular layers, giving rise to a local distortion of the lattice. Sometimes an impurity may be present; a certain proportion of the sites which should ideally contain chloride ions may in fact contain bromide, or even sulphate, ions. These various defects have important practical consequences. The very fact that crystals grow at all seems to depend on certain structural faults of this kind. Other consequences are enhanced mechanical strength, electrical conductivity, catalytic powers, photochemical properties such as photoconductance, phosphorescence and photosensitivity, and the presence of anomalous colour, e.g. blue calcium fluoride. For our present purpose the interest of such defects is that they must affect the composition of the material. A crystal which has the ideal composition A : B = 1 : 1 (say), but which deviates from the ideal, is said to be *non-stoicheiometric*. The phenomenon is very common. In most cases, to be sure, the deviation is too slight to be detected by chemical analysis; rather has it to be inferred from some of the special consequences listed above. But, quite often the deviation is large enough to affect the composition seriously. For instance, the compound, wüstite, conventionally represented by the formula FeO, never has this composition. Different specimens vary in composition, and we may write $FeO_{(1+x)}$, where x is 0·06–0·19. The explanation is that, whilst the crystal sites allocated to O^{2-} ions are fully occupied, some of those allocated to Fe^{2+} are vacant; and that, in order to maintain electrical neutrality, some of these cationic sites contain Fe^{3+} ions instead. For example, in a specimen with the composition $FeO_{1·10}$, there would be, for every eleven O^{2-} ions, eight iron sites containing Fe^{2+}, two containing Fe^{3+}, and one empty. It is impossible to attribute a simple integral quantivalence to the iron in such a compound.

7.5 Fractional covalency numbers

We might then try to avoid some of these difficulties by restricting the use of quantivalence to compounds with clearly defined molecules. Even so, other difficulties are liable to appear. Let us consider a simple conjugated molecule such as that of 1:3-butadiene (1). In v.b. terminology this molecule is regarded as a resonance hybrid

$$H_2C=CH-CH=CH_2 \quad H_2\overset{\cdots\cdots\cdots\cdots}{\overset{\cdot}{C}-CH=CH-\overset{\cdot}{C}H_2}$$

$$(1) \qquad\qquad\qquad (2)$$

$$H_2\overset{\ominus}{C}-CH=CH-\overset{\oplus}{C}H_2$$

$$(3)$$

with (1) as the principal form, but with lesser contributions from (2), and from (3) and the corresponding other form with reversed polarity. (The two lone electrons on the terminal atoms of (2) have their spins paired, so that a 'formal bond'—indicated by the broken line—exists between them.) So long as we keep to the v.b. approximation, the quantivalence of carbon remains four; this is inherent in the method, provided we count the formal bond in formula (2) and the bond implied by the formal charges in (3) as valencies. On the other hand, the molecular orbital method leads to a different result. Calculations show that the π-bond order between the centre carbon atoms is 0·45, and that between one of these inner atoms and the outer one is 0·89. Counting the order of the σ-bonds, whether between carbon and carbon or between carbon and hydrogen, as unity, the total valency of a terminal carbon atom comes out as $3 + 0·89 = 3·89$, whilst that of an inner one is $3 + 0·89 + 0·45 = 4·34$. In a similar way the total quantivalence of each carbon atom in benzene is reckoned to be $3 + 2 \times 0{\cdot}67 = 4{\cdot}34$, by molecular orbital calculation.

7.6   Limitations of the bond concept

The idea of a definite quantivalence has a corollary in the idea of definite bonds, whether covalent or ionic. With the recognition of a certain vagueness in the one idea, a corresponding vagueness must affect the other also. By hypothesis, the v.b. method makes use of palpable linkages, but the m.o. method does not necessarily. In particular, when non-localized m.o. are involved, we cannot speak of a bond at all in any ordinary sense. The accommodating of the six electrons in three of the aromatic π-orbitals of benzene results in a diminution of energy, and consequently to a tighter binding of the six carbon atoms. But no definite bonds can be drawn.

The hybrides of boron are difficult to represent by electronic bond-diagrams, as we saw in §**3.8**. Formula (21) of that section shows the structure found experimentally for the molecule of the simplest such hybride, B_2H_6. It involves a double hydrogen bridge across the centre. Two types of electronic interpretation have been proposed: (a) that there is a new type of hydrogen bonding represented as resonance between (4) and (5); (b) that the terminal BH_2-groups are held together by two 'three-centre bonds', each such bond—indicated by a broken line in (6)—consisting of an electron pair accommodated in a bonding molecular order covering the three atoms B, H and B.

(4)   (5)   (6)

The former is of the v.b. type, the latter of the m.o. type (and several variants of it have been suggested). A number of boron compounds show this anomaly, and it is clear that their molecules cannot be simply represented with conventional bonds only.

Over 120 years ago Zeise discovered a series of coordination compounds containing ethylene; for instance—in modern formulation —$K[PtCl_3C_2H_4].H_2O$. It is not easy to explain how the ethylene molecule is attached to the platinum atom. X-Ray analysis now shows that the chlorines are situated at three corners of a square, with the platinum atom at its centre, and that the ethylene molecule has its mid-point at the fourth corner, the carbon–carbon bond being perpendicular to the plane of the square. Infra-red evidence indicates that the ethylenic bond is not greatly modified. The electronic mechanism which seems to account for these and other properties is as follows: the platinum atom uses dsp^2-hybridized orbitals (see §5.5) three of them accepting electron pairs from the chlorines; the ethylene is partly bound by use of a m.o. formed by overlap between the fourth dsp^2-orbital of the platinum and the π-orbital of the double bond; this bonding is reinforced because two other (occupied) orbitals of the platinum atom are in a position where they can overlap with an unoccupied orbital of each carbon. Neither of these effects would alone confer sufficient stability; to secure the ethylene, the cooperation of both is needed, and this is favoured by the stereochemical situation. Many π-bonded complexes of this sort are now known. Once more the simple bond concept fails.

7.7 Quantivalence imposed by geometrical considerations

We have seen that both covalency and electrovalency have a mechanism of valency saturation, though different ones. The various van der Waals forces on the other hand—except the hydrogen bond—have less inherent need to show saturation. If an ion attracts a dipolar molecule, its power to attract a second is not seriously affected. Compounds resulting from these forces are therefore not under the same direct necessity of having stoicheiometric compositions as are other types. But a tendency towards some definite composition may be imposed by geometry.

The *interstitial compounds* formed by the transitional elements illustrate one possibility. These metals usually crystallize with their atoms *close-packed*. Spheres cannot fill space completely; between them, when they are packed together as closely as possible, there are interstices—or 'holes'—of two kinds: tetrahedral, between four spheres, and octahedral, between six; for every sphere (or atom) there are two of the former and one of the latter. Now it happens that these holes are large enough to accommodate small atoms, such as hydrogen or carbon or nitrogen, with only a slight distortion of the lattice, or even with none. The familiar solubility of hydrogen in palladium arises in this way. The energy needed to dissociate the hydrogen molecule into atoms is supplied by the van der Waals and metallic forces exerted on the interstitial atoms. The amount of hydrogen absorbed can vary over a wide range, and so the product is usually regarded as a solid *solution*. But, if all the interstices are occupied—or a definite fraction of them—then the product will have an integral composition and can be called a compound. A hard-and-fast distinction cannot always be made; but, when the composition is definite enough to warrant the use of the word 'compound' this is due to the geometrical factor. In the compound TiH_2, for example, the hydrogen atoms occupy all the tetrahedral holes between the titanium atoms; whilst in the compound W_2C (which is one of the hardest substances known), the carbon atoms occupy half the octahedral holes.

In this type of compound, the interstices between the metal atoms can accommodate only very small atoms. By contrast, in a series of compounds recently studied, there may be 'cages' in which quite large molecules can be entrapped. Such cages arise when one constituent of the compound is hydrogen-bonded in such a way as to lead to a very open structure. Best known are the *clathrate compounds* of Powell. Molecular compounds between quinol (Q) and another substance (M) had been known for many years, and they generally have the composition, Q_3M. An example with H_2S for M had been described by Wöhler in 1849; and amongst other possibilities are SO_2, methanol, methyl cyanide and argon. Powell showed that in all of them the quinol adopts an alternative crystal structure wherein the quinol molecules ($C_6H_4(OH)_2$) are hydrogen-bonded to one another, giving a cage large enough to hold a molecule of radius about 2 Å for every three quinol molecules, and that it is here that the second constituent is in fact situated. Almost any molecule can be taken up, provided it is not too large in size or too awkward in shape. The composition, Q_3M, is imposed simply by geometrical considerations. In fact it is usually found that there is rather less of M present than the formula requires. A certain proportion of the available sites is unoccupied.

Another type of clathrate compound occurs in the 'inert gas hydrates'. These are compounds of water with some other substance, which may be an inert gas in the literal sense, but may also be chlorine, methyl chloride, etc. (The chlorine is then 'inert' only in that it is not chemically combined in the ordinary sense.) In these the water has crystallized in a modified ice structure (cubic) containing various sizes of cages between the hydrogen-bonded water molecules. The 'inert gas' molecules are contained in the larger of these cages. In one example there are six cages large enough to trap a chlorine molecule for every 46 water molecules. Hence we have $6Cl_2.46H_2O$, formerly supposed to be $Cl_2.8H_2O$.

Urea also crystallizes with a hydrogen-bonded structure. Normally this is in the tetragonal system; but, if during crystallization from water long-chain aliphatic compounds are present, a modified, hexagonal pattern is adopted. In this the urea molecules act rather like the wax walls of a honeycomb, long channels of radius about $2\frac{1}{2}$ Å being left vacant; along these lie extended the aliphatic molecules, corresponding to the honey. Since only suitably slim molecules will fit into these channels, a method was developed for separating paraffin-like compounds of different types, and it has been used industrially. Straight-chained molecules will give urea-adducts; branched-chained molecules will not. These compounds could be described as stoicheiometric in two dimensions, but not necessarily so in the third. The number of channels per unit cell is two; but how many paraffin molecules are taken up per cell depends on their length. With hexane the composition of the adduct is about, hydrocarbon:urea = $1:6\cdot2$; with decane about $1:8\cdot5$.

The blue coloured starch iodide has a similar structure. The carbohydrate units are linked so as to leave hexagonal channels, along which lie the I_2 molecules.

7.8   Conclusion of the matter

The theme of this chapter has been that the straightforward notions of quantivalence, of the molecule, and of the bond, are not always applicable to the complex facts of chemistry. Nevertheless they have been very useful in the past, and they will continue to be useful especially if we realize their limitations. It has been wisely written that 'the pursuit of exact verbal definitions . . . is rarely of much scientific value, provided the requirements of convenience and consistency are satisfied'.

Problems

1. Discuss the electronic formulation, the shape and the dimensions of the molecules, NO_2^+, NO_2 and NO_2^-.
2. Draw a scale-diagram of the molecule of 1:2-dibromoethylene, for both *cis*- and *trans*-isomers. Hence estimate the distances between the Br atoms. Which molecule would have a non-zero dipole moment?
3. Write the four Kekulé formulae for the molecule of anthracene. Hence estimate the bond-orders of the structurally different bonds. Use the graph in Fig. 5.7 to predict probable bond-lengths.
4. Which of the following molecules would you expect to be (a) linear, (b) planar? H_2S, PH_3, N_2O, BCl_3, N_3^- and ClO_2.
5. Estimate probable dimensions for the following molecules: CBr_2H_2, $Cl—C\equiv C—H$, $H—C\equiv C—C\equiv N$, *m*-dichlorobenzene.
6. Which of the following molecules would you expect to have a zero dipole moment? *p*-dinitrobenzene, phenol, *p*-dihydroxybenzene, 1:2-dichloro-ethane.
7. If the heat $(-\Delta H)$ of formation of gaseous naphthalene is -34 kcal/mole, use the data in §4.13 to find its heat of formation from atoms. Hence assess a value for the resonance energy.
8. Use the following values of the dissociation constant of acetic acid dimer, in benzene solution, to obtain a value for the dissociation energy of the hydrogen bond:

Temperature (°C)	6·27	25·00	45·08
K_D (l./mole × 10^3)	2·67	7·48	19·40

9. Spectroscopic measurements show that the CO_2-molecule has a single moment of inertia of $71·67 \times 10^{-40}$ c.g.s. unit. Show this to be consistent with a linear molecule having C—O distances of 1·16 Å.
10. Use the data in Table 4.5 to find heat of hydrogenation of ethylene. Use the result to estimate a value for the resonance energy of benzene, given that its heat of hydrogenation, in the gaseous state, is about 50 kcal/mole.
11. Write as many bond-diagrams as you can for the odd molecules CR_3, where R is (a) C_6H_5, and (b) 2:4:6-trinitrophenyl.
12. The heats of combustion $(-\Delta H)$ of ethylene and ethane are 337 and 373 kcal/mole, and those of toluene and of methylcyclohexane are 944 and 1100 kcal/mole. What is the resonance (or delocalization) energy associated with the aromatic ring in toluene?
13. Give explanations in valency bond terminology of the following observations: (a) The central C—C bond in 1:3-butadiene has a length of 1·46 Å. (b) The pK-value of acetic acid is about 4·8, whilst that of ethanol

is about 18. (c) The N—O distances in the nitrobenzene molecule are each about 1·21 Å.

14. Accurate measurements reveal a significant variation in the following C—H distances: 1·09 Å in methane; 1·08 in benzene; 1·07 in ethylene and 1·06 in acetylene. Interpret the variation with particular reference to the type of hybridization occurring in these molecules.

15. How long would it take an electron, accelerated by a potential difference of 40,000 V, to travel the length of a Cl_2-molecule? Roughly how many (stretching) vibrations would the molecule make in this time?

16. The solvent power of water for salts has been explained in two ways: as due (a) to the solvation of the ions, which liberates sufficient energy to compensate for the energy needed to overcome the strong attraction between the ions in a crystal; and (b) to its high dielectric constant, which diminishes the attraction between the ions in solution. Explain why these two explanations are closely related.

17. The valency angle in the H_2S-molecule is about 92°. Use this datum, along with the electronegativities of sulphur and hydrogen, to estimate a value for the molecular dipole moment.

Index

(Molecules are indexed under their chemical formulae)